Smart Cities

Smart city development has emerged as a major issue over the past five years. Since the launch of IBM's Smart Planet and Cisco's Smart Cities and Communities programmes, their potential to deliver on global sustainable development targets has captured the public's attention. However, despite this growing interest in the development of smart cities, little has as yet been published that either sets out the state of-the-art, or which offers a less than subjective, arm's length and dispassionate account of their potential contribution.

This book brings together cutting-edge research and the findings of technical development projects from leading authorities within the field to capture the transition to smart cities. It explores what is understood about smart cities, paying particular attention to the governance, modelling and analysis of the transition that smart cities seek to represent. In paving the way for such a representation, the book starts to account for the social capital of smart communities and begins the task of modelling their embedded intelligence through an analysis of what the 'embedded intelligence of smart cities' contributes to the sustainability of urban development.

This innovative book offers an interdisciplinary perspective and shall be of interest to researchers, policy analysts and technical experts involved in and responsible for the planning, development and design of smart cities. It will also be of particular value to final year undergraduate and postgraduate students interested in Geography, Architecture and Planning.

Mark Deakin is Professor of Built Environment in the School of Engineering and Built Environment, Edinburgh Napier University. He is also Head of the Centre for Sustainable Communities in the Institute for Sustainable Construction, at Edinburgh Napier University. His research focuses on sustainable urban development, intelligent cities, smart cities and communities.

Smart Cities

Governing, modelling and analysing the transition

Edited by
Mark Deakin

Routledge
Taylor & Francis Group

LONDON AND NEW YORK

First published 2014
by Routledge
2 Park Square, Milton Park, Abingdon, Oxfordshire OX14 4RN

Simultaneously published in the USA and Canada
by Routledge
711 Third Avenue, New York, NY 10017

First issued in paperback 2015

Routledge is an imprint of the Taylor & Francis Group, an informa business

British Library Cataloguing in Publication Data
A catalogue record for this book is available from the British Library

Library of Congress Cataloguing in Publication Data
  Smart cities : governing, modelling, and analysing the transition / [edited by] Mark Deakin.
    pages cm
  1. City planning. 2. City planning – Technological innovations. 3. Cities and
  towns – Growth. 4. Sustainable development. I. Deakin, Mark editor of compilation.
  HT166.S5877 2013
  307.1'216 – dc23   2013002744

ISBN13: 978-1-138-93272-2 (pbk)
ISBN13: 978-0-415-65819-5 (hbk)

Typeset in Times New Roman
by Out of House Publishing

Contents

Figures

Tables

Contributors

Andrea Caragliu, Politecnico di Milano, Piazza Leonardo 32, 20133 Milan, Italy

Ian Cooper, Eclipse Research Consultants, 121 Arbury Road, Cambridge, UK, CB4 2JD

Peter Cruickshank, School of Computing, Edinburgh Napier University, Edinburgh, Scotland, EH10 5DT

Mark Deakin, School of Engineering and Built Environment, Edinburgh Napier University, Edinburgh, Scotland, EH10 5DT

Chiara Del Bo, Università degli Studi di Milano, Via Conservatorio 7, 20122 Milan, Italy

Silvia Giordano, Dipartimento Casa-Città, Politecnico di Torino, V.le Mattioli 39, 10125 Torino, Italy

Nicos Komninos, URENIO Research, Aristotle University of Thessaloniki, Greece

Karima Kourtit, VU University, De Boelelaan 1105, Amsterdam 1081 HV, The Netherlands

Loet Leydesdorff, Amsterdam School of Communication Research (ASCoR), University of Amsterdam, Kloveniersburgwal 48, 1012 CX Amsterdam, The Netherlands

Patrizia Lombardi, Dipartimento Casa-Città, Politecnico di Torino, V.le Mattioli 39, 10125 Torino, Italy

Peter Nijkamp, VU University, De Boelelaan 1105, Amsterdam 1081 HV, The Netherlands

Krassimira Paskaleva, Herbert Simon Institute, 8.30 Harold Hankins Building, Manchester Business School, Booth Street West, Manchester, UK, M15 6PB

1 Introduction (to smart cities)

Mark Deakin

In a recent article, Hollands (2008) asks the question: 'Will the real smart city stand up?' For according to Hollands, cities all too often claim to be smart, but do so without defining what this means, or offering any evidence to support such proclamations. The all too often 'self-congratulatory' tone cities strike when making such claims does not seem to sit well with Hollands (2008). For while images of the digital city, intelligent city, high-tech district and neighbourhoods of smart communities abound, they all fail to convey what it means to be smart and why it is important for cities to be defined in such terms.

In Hollands' (2008) opinion, the validity of any city's claim to be smart has to be based on something more than their use of information and communication technologies (ICTs). Hollands asks this question because cities all over the world are beginning to do just this and use such technologies as a means of branding themselves smart. Such smart city forerunners as San Diego, San Francisco, Ottawa, Brisbane, Amsterdam, Kyoto and Bangalore are all now setting a trend for others to follow. The other cities keen to follow in their wake and become smart include: Southampton, Manchester, Newcastle, Edinburgh, Edmonton, Vancouver and Montreal. It appears the rush to become a smart city has begun to gather apace and, as a consequence, pressure is now growing for cities to become even smarter.

Taking Hollands' (2008) paper on the transition from intelligent to smart cities as its point of departure, this book takes the opportunity to reflect upon the anxieties currently surrounding the governance, models and analysis of such developments.

In this aim, Chapter 2 reflects upon some of the anxieties surrounding the transition from intelligent to smart cities drawn attention to by Hollands (2008). In particular, the anxiety that it has more to do with cities meeting the needs of the market than the intelligence that is required for them to be smart. Working on the assumption that any attempt to overcome such an anxiety means shifting attention away from the needs of the market and towards the intelligence which is required for cities to be smart, this chapter begins to set out a less presumptuous, more critically aware and insightful understanding of the transition from intelligent to smart cities. For the representation of smart cities this chapter advances is founded on the realisation that it is the legacy of work undertaken on the informational basis of communications embedded in the very notion of intelligence, which is critical to understanding what it means for cities to be smart.

What follows goes on to capture the information-rich and highly communicative qualities of the technical, social, wider environmental and cultural intelligence currently surrounding the transition to smart cities. In particular the acute methodological issues they pose and critically insightful role which the networks of innovation and creative partnerships underlying these developments play in the learning, knowledge transfer and capacity-building exercises that service the transition to smart cities. This is what the chapter suggests Hollands' (2008) account of smart cities misses and goes some way to explain why he asks 'the real smart city to please stand up!' For in cutting across the legacy of the transition from the informational, to the intelligent and now smart city, Hollands' (2008) account of this transformation is not as well grounded in the informational and communicative qualities of the embedded intelligence they are built on and stand for.

This, the chapter suggests, is a critical insight of some note, for only in giving such a well-grounded account of the embedded intelligence drawn attention to, is it possible to do what Hollands (2008) asks of the transition. That is, 'under-gird' the social capital, which is not only critical in underpinning the informational and communicative qualities of the embedded intelligence that smart cities stand on, but pivotal in gaining a fuller insight into their wider environmental and cultural significance. In particular, in gaining a fuller insight into the wider environmental and cultural significance their networks of innovation and creative partnerships take on in embedding the intelligence of such an informatics-based and community-led transition to smart cities.

The chapter goes on to suggest this insight is equally significant for the reason it takes Hollands' (2008) thinking full circle. That is to say, by offering an alternative to the very 'top-down' entrepreneurial based business logic which is called for. Something which this chapter in turn realises by turning the top-down entrepreneurial based business logic of existing accounts on their head and grounding the information-rich and highly communicative qualities of such environmental and cultural developments on the community-led logic that is emerging to support the transition. That logic which in real time is aligned with, not against, the cybernetics of the social capital underlying the environmental and cultural significance of smart communities and what their founding networks of innovation and creative partnerships embed as the intelligence of this transition. Those developments in the cybernetics of social capital whose emergent environmental and cultural qualities are currently in the process of being institutionalised in the learning, knowledge transfer and capacity-building exercises which are intelligent in embedding the informatics of this community-led transition to smart cities.

Chapter 3 examines how electronic governance (e-governance) can assist in helping cities make better decisions and become more competitive, as well as engage with citizens in democratic activities and decision-making processes. This chapter suggests that e-governance may provide a new vehicle for the emerging smart city. It assesses the conceptual landscape for city e-governance, and focuses on how cities can foster collaborative digital environments to enable local competitiveness and prosperity through knowledge networks and part-nerships, integrated e-services and e-participation. Drawing from the literature and the results of a comprehensive survey study in 12 European cities, it puts forward a series of propositions on the future of e-city governance in Europe and the implications for strategic policy innovations to foster smart cities.

While Chapters 2 and 3 review the debate on governing the transition to smart cities, Chapters 4, 5, 6 and 7 examine the transition from intelligent cities to smart cities. In this respect, Chapter 4 develops the notion of an intelligent city as the provider of electronically enhanced services. It identifies how this growing interest in the notion of intelligent cities has led univer-sities to explore the possibilities of using communities of practice (CoPs) as a means of drawing upon the industrial knowledge base such organisations offer to develop integrated models of e-government (eGov) services. It reports on the attempts made by a consortium of leading European cities (led by Manchester) to use the intelligence that CoPs generate as the environmental and cultural means by which to work smarter in developing integrated models of eGov service provision.

Made up of researchers, computer engineers, informational managers and public sector service providers, the IntelCities CoP has worked to develop an integrated model of eGov services and support the actions taken by cities to host them on platforms with sufficient intelligence to meet the e-learning needs, knowledge transfer requirements and capacity building commitments of their socially inclusive and participatory urban regeneration programmes.

As an exercise in CoP development, this organisation is particularly suc-cessful for the reason the intelligence it has sought to embed in cities and integrate within their platforms of eGov services is inter-organisational, net-worked, virtual and managed as part of a highly distributed web-based learn-ing environment. Made up of both open source software groups, experts and lay people, the CoP is culturally unique in the sense its network provides an example of a virtual organisation set up to manage the learning needs and knowledge requirements of a technological platform. As such it:

- offers the means to meet the learning needs, knowledge transfer require-ments and capacity building commitments of the organisation;
- co-designs them as a set of services that are socially inclusive and par-ticipatory and which allow users to learn about the availability of such services, how to access them and the opportunities they offer everyone to become engaged with and get involved in meeting the knowledge transfer requirements and capacity building commitments of their urban regen-eration programmes;
- allows for the monitoring and evaluation of such actions.

As the chapter establishes, it is the e-learning platform that makes it possible for the online services under development to be integrated with the knowledge transfer and capacity-building technologies which are needed for this CoP to work as a shared enterprise. That is as an enterprise which allows such organisations to collaborate and build consensus on the competencies, skills and training that are needed to service the required online developments.

Together, the networks, innovation and creativity of the partnerships responsible for organising the development of these technologies, skills and training exercises make it possible to engage citizens and show how the active participation of communities is not only intelligent, but smart. This turns attention to what is termed the eTopia demonstrator, developed to illustrate the functionality of the semantically rich eGov services in question. This term is borrowed from the account of intelligent cities as e-topias and as organisations that are SMART, lean, mean, green software systems, driven by networked communities which are virtual. Organisational characteristics that are themselves built on the learning needs, knowledge management requirements and digital libraries of the electronically enhanced services that are made available on the eCity platform as a pool of integrated eGov services.

Chapter 5 goes on to examine the spatial intelligence of cities, the use of digital technologies and the institutional settings of those innovation systems seen as smart enough to radically transform cities. The starting point for this chapter are two related observations about the increased use of terms like intelligent and smart in contemporary urban planning and development. The first concerns the somewhat over-simplistic way cities tend to use the terms 'intelligent' or 'smart'. The second relates to the diverse range of strategies cities are currently assembling in laying claim to such a status. The observation here being that such a diverse range of strategies tends to say more about the ambiguity of the relationship digital technologies have to the developments under examination, than what it means for them to be ether intelligent or smart. This is because for Komninos, the strategies in question are seen as being insufficiently developed for their digital technologies to embed the intelligence cities need to be smart and therefore require to claim such a status.

As a counterpoise to these observations, this chapter lays down some of the 'fundamentals of spatial intelligence', whose strategies and applications can be seen as being smart. It argues that despite the great diversity of strategies and applications, the logistics of spatial intelligence teach us that smart cities rest on a few knowledge-based trajectories. In particular, those knowledge-based trajectories that are embedded in the transitions of Bletchley Park, Hong Kong and Amsterdam and which the paper suggests are still only partially understood.

Taking Komninos' idea that cities are still stuck in the digital, rather than embedded in the intelligence of what is smart, as the 'third' observation on the transition, Chapter 6 examines the thesis on the 'embedded intelligence of smart cities'. For, as the chapter points out, while Mitchell (1995) sets out a vision of urban life literally done to bits, left fragmented and in danger of coming unstuck, 'e-topia' offers a counter-point to this and image of the city no longer left in bits, but a place 'where it all comes together'.

Dwelling on the reconciliatory nature of these symbolic statements, this chapter suggests that while this thesis on the 'coming together' of the virtual and physical and dissolution of the boundaries between 'cyber and meat space' is compelling, there are a number of concerns surrounding the technical, social and environmental status of the embedded intelligence that is currently available for urban planners and developers to make cities smart. While problematic in itself, the chapter also suggests that if the difficulties experienced over the transition from intelligent to smart cities were only methodological they may perhaps be manageable, but the problem is they run deeper than this and relate to more substantive issues which surround the cultural trajectory of the thesis.

This, the chapter suggests, is a critical insight of social and environmental significance because if the cultural trajectory of the thesis is not in the direction of either the embedded intelligence of smart cities, or the ICTs of what is referred to as 'digitally inclusive regeneration platforms', then the question arises as to whether the whole notion of e-topia can be seen as a progressive force for change, or merely a way for the embedded intelligence of smart cities to reproduce the status quo.

This unfortunate scenario is what Graham and Marvin (2001) refer to not as e-topia but as splintering urbanism, because, under this thesis, the citizenship underlying the informatics of these communities is no longer able to carry the sheer weight of the material that such a cybernetic-based networking of intelligence is supposed to support. This, the chapter suggests, is important because such a representation of the transition offers what can only be referred to as the antithesis to e-topia. An antithesis that, it might well be added, goes some length to search out, uncover and expose the other side of this cybernetic-based intelligence and reveal what currently lies hidden in the debate that is currently taking place about the transition to smart cities.

From this perspective, the chapter suggests that it is evident that the problems with e-topia are as much substantive as methodological, the former holding the key to the latter. In substantive terms the chapter offers another twist on the question of what the transition from intelligent to smart cities means and in doing so goes very much against the grain, arguing that our current understanding of embedded intelligence, smart cities and the ICTs of digitally inclusive regeneration puts us on the verge of a new environmental determinism. An environmental determinism that this time around is cybernetic, in that it is founded on the embedded intelligence of knowledge-based agents underpinning the networking of smart cities and the digitally inclusive regeneration platforms they support.

To avoid repeating this mistake (yet) again, attention is drawn to the spaces that their radical democratic, i.e. egalitarian and ecologically integral, account of the transition opens up for a much more emancipatory view of the intelligence embedded in those knowledge-based agents understood as being smart enough to meet these requirements. Those knowledge-based agents, it should perhaps be added, understood as being smart enough to meet these requirements by way of and through their exploitation of the social capital that underlies the very communities that give rise to the environmental and cultural norms, rules and values of such developments. In particular, the social capital that underlies the embedded intelligence of smart cities and that their communities of knowledge-based agents (architects, planners, engineers and surveyors) in turn support by hosting them as environmental and cultural services found on digitally inclusive regeneration platforms.

This chapter suggests that in ignoring these warnings and being unable to learn the lessons that such a critical reworking of the thesis offers, the strategy previously adopted must be seen as suspect. Not only because the vision and scenarios it advances have a tendency to side-step the social significance of digital technologies, but for the reason that in doing so, they end up replacing the agonies of equality and ecological integrity with little more than the 'gnostics' of 'new age' environmental and cultural storylines centred around the quality of life. The strategy advocated for adoption by this chapter is not grounded in such rhetoric.

The vision of e-topia this chapter builds instead rests on the messages others advance by turning the tables and agreeing that, while words offer the possibility of 'bringing what it all means back together', actually turning things around lies not so much in the words as with the semantics of the syntax and vocabulary governing the digitally inclusive nature of the regenerative storylines emerging from this discourse and, perhaps even more importantly, the degree to which they overcome the divided antagonisms of the excluded. This way, the chapter suggests, it becomes possible for the multiplied memory and infinite mind of the 'cyborg civics' and environments of their 'tribe-like culture', not to so much bemoan the 'nomadicity of wireless bi-peds', but actively celebrate the creativity of the virtual communities emerging from the digital-inclusive nature of such regenerative storylines.

In particular, it is added, celebrate the opportunity that this in turn creates for virtual communities to use the collective memory, wikis and blogs of their electronically enhanced services as a means for such platforms to bridge social divisions. Bridge them – it is important to note – by drawing upon the political subjectivities of cyborg-civics, their environment, tribe-like culture and nomadicity, as wireless bi-peds with the embedded intelligence smart enough for the citizens of this community to span them. In particular, span them with bridges that are not merely symbolic, but real in the sense that the semantic web of this knowledge base serves to be the agent of something more than a prop. Something more than a prop and stronger in the sense which the embedding of such intelligence allows the wikis and blogs of the web-based services that supports all of this to begin doing the job asked of them. That is the job of building a stage which is large enough for the analytic, synthetic and symbolic components of the transition to be smart in playing out the possibilities there are for urban planning to be both equitable and ecologically integral.

Chapter 7 suggests that over the course of the past decade, the smart cities agenda is an issue that has gained real momentum in the Europe. The significance of this being reinforced further by other international organisations, such as the OECD, which suggests that smart cities offer society the prospect of not only being environmentally sustainable, but sufficiently competitive and cohesive for their emerging culture to meet their pressing quality of life agenda. As the chapter points out, as a result of such high-ranking institutional support, many cities have now adopted this socially cohesive, environmentally sound and economically competitive reading of what it means to be smart as a way to profile themselves as forward-looking, prosperous and well-endowed cultures. For instance:

- the Amsterdam smart city initiative emphasises the importance of collaboration between the citizens, government and businesses to develop smart projects that will 'change the world' by saving energy;
- Southampton City Council uses smart cards to stress the importance of integrated e-services;
- the City of Edinburgh Council has formed a smart city vision around an action plan for government transformation;
- the Malta smart city strategy promotes a business park to achieve more economic growth;
- IBM, Siemens and ORACLE have formed their visions of the smart planet;
- a number of EU research projects have also to deal with various issues of the smart city. A recently concluded pan-European research project, IntelCities, for example, concluded that governance, as a process and outcome of joint decision-making, has a leading role to play in building the smart city, and that cities should develop collaborative digital environments to boost local competitiveness and prosperity by using knowledge networks and partnerships, integrated e-services and governance.[1]
- the INTERREG project Smart Cities is using an innovation network between university, industrial and governmental partners to develop the triple helix of e-services in the North Sea Region by cultivating a novel customisation process.

It is this view of smart-er cities as people-based, human, socially inclusive, environmentally sensitive and culturally aware that the chapter advocates. Adopting this 'digitally inclusive' vision of cities, the chapter reflects upon the current trends and understanding of what it means for urban administrations, policy makers and businesses in Europe to be smart. In developing this vision, the chapter pays particular attention to the role of the smart city as a nexus for open innovation and how the strategic significance of this development has become the mainstay of current discussions about the Future Internet, living labs, innovation and competitiveness-driven (urban) development.

By conducting a critical review of some high-profile programmes and initiatives on smart cities, the emerging trends are explored and insights are drawn about the challenges these developments pose. The said analysis is based on four smart city projects and their relevant EU programmes. They are chosen because collectively they reveal what Europe expects smart cities to stand for.

Chapter 8 turns the table on these technical accounts of smart cities and examines the developments from the neo-evolutionary perspective of the triple helix. Going against the grain, this chapter suggests the reinvention of cities currently taking place under the so-called 'urban renaissance' cannot be defined as a top-level 'trans-disciplinary' issue without a considerable amount of cultural reconstruction. It suggests the highly distributed character of this reconstruction has not yet been given the consideration it demands.

For the authors, accounts of this cultural reconstruction have tended to reify the global perspective and fail to appreciate the meta-stabilising transformation of innovations systematically worked out as the informational content of 'social and cultural processes' operating at the local level. They also suggest that the potential this meta-stabilising mechanism has to work as a reflexive layer is what lies behind the surge of academic interest currently focusing on communities as the practical manifestation of intellectual capital and means to exploit the knowledge generated from their organisation.

For them the significance of this 'knowledge-based reconstruction' is seen to rest on the real capacity such an environment has to not only be cultural but economic, insofar as such a reinvention of cities makes it possible for urban regeneration programmes to function as systems of innovation that respond to the 'creative destruction' of the global and 'reflexive reconstruction' of the local settings that they relate to. They stress the 'creative reflexivity' of this meta-stabilisation is far from 'symbolic', or of merely representational significance, in the sense that it generates the critical reinforcement that civil society needs to govern any such 'programmatic' integration of cities into their emerging innovation systems.

Chapter 9 draws upon the representation of smart cities, intelligent enterprise architectures and business models underlying the triple helix, as a means to demonstrate how this model can be put to good use as a knowledge base for universities, industry and government to draw upon. This turns the representational logic of the previous chapters on its head by beginning with the SmartCities (inter) Regional Academic Network's (SCRAN's) three-way partnership and this organisation's way of working, i.e. methodology. Starting with the triple helix of SCRAN's methodology, this chapter draws particular attention to how the communication needs and technical requirements of such a three-way partnership can be met. In particular, how in methodological terms they can be met in ways seen, understood and known to be smart.

In setting out how the organisation in question is doing this, the paper goes on to configure the triple helix of SCRAN and set out the 'step-wise' logic underlying the partnership's emergent knowledge base. Having done this, attention turns towards the networking of the triple helix by the universities and industrial sectors. From here attention turns towards the methodological question of how this knowledge base can be used as a learning platform for the partnership to take government-led programmes of electronically enhanced service developments full circle.

The chapter suggests it is the three-way partnership between universities, industry and government that captures the science and technology around which this triple helix of regional development turns. This goes some way to capture SCRAN's particular take on the triple helix and serves as a means of drawing attention to the science and technology underpinning the strategic research funded by the EC to support the innovative and creative use of ICTs as a platform for the electronic enhancement of government services. This in turn serves to highlight the innovative and creative capacities of the SmartCities partnership and what can perhaps best be referred to as a statement of the network's eGov services capability. From here the organisational means needed for business to meet these standards and those also required by government to satisfy their expectations are explored in terms of the partnership's ability to begin co-designing the (inter)regional development of eGov services and do so in a way that allows the transnational dimensions of this programme to be mainstreamed across the North Sea.

This representation of the triple helix goes some way to uncover the scientific and technical capacity of the research-based services that the network offers. In particular, what the model offers the SmartCities partnership as a platform of ICTs supporting:

- the policies, plans, programmes and projects underlying the (inter) regional development of electronically enhanced government services that this knowledge is based on;
- the legacy of experiential learning that the network brings to the project;
- the management of such vehicles as a common resource, shared with other (business and government sector) agencies, institutions and contacts within the partnership;
- the competencies that the network needs to distribute across the partnership;
- governments' drive for efficiency, joined-up working, reduced bureaucracy, back-office reorganisation and customer focus, as vehicles for delivering on policy commitments made by administrations on their corporate performance, economic competitiveness and social cohesion.

Under the heading of capacity building, this chapter goes on to propose that organisations like SCRAN should focus attention on the underlying technical issues and, in particular, those that support the partnership's enterprise architecture and business models. The knowledge infrastructure chosen by the organisation to support these developments is then outlined as a means to provide a critical insight into how this technology can be used to answer questions raised about the cultural and environmental attributes of smart cities.

The chapter suggests this shift towards the triple helix of smart cities among policy makers, researchers and members of the business community is significant because it offers the opportunity for regional innovation to replace so-called mode 2 notions of policy, research and technical development, i.e. hierarchies and networks, with CoPs. In particular, with CoPs that offer the intellectual capital needed for the innovation systems that cities are currently assembling to be smart in creating wealth from the regional advantage their ICT-related developments in turn construct.

Chapter 10 provides the evidence base to support many of the previous statements made about the critical nature of the relationship between intellectual capital, wealth creation and ICTs of government-led developments. In line with the triple helix, it begins with the assumption that the intellectual capital of wealth creation depends not only on the endowment of hard infrastructure ('physical capital'), but also on the social, cultural and environmental capital of the city's knowledge base.

This original and unique analysis provides the statistical basis to examine the relationship between intellectual capital, wealth creation and development of government services by smart cities across Europe. Using data drawn from the EC's Urban Audit statistics, the chapter attempts to analyse the factors determining the development of smart cities. It highlights the presence of a 'creative class', with the intellectual capital to access high-quality urban assets, educational opportunities and government services, underlying the development of smart cities. More significantly, it identifies the intellectual capital allowing access to such assets, opportunities and services, which are all positively correlated with the creation of wealth. The findings are important because they start to build the evidence base to support the previously unchallenged assumption about the existence of a positive relationship between the pool of intellectual capital, stock of wealth created and their governance.

Given the wide-ranging and generic nature of these findings, the chapter then outlines how this representation of the transition can be used to bootstrap the technology of innovation systems and exploit their intellectual capital to generate wealth. The outcome of this examination is then used to formulate a new strategic agenda for governing the development of smart cities across Europe.

Chapter 11 offers an advanced triple helix model of smart cities, built around SCRAN and the development of an analytical framework able to verify whether the performance of cities is smart. It offers a profound analysis of the interrelations between the components of smart cities, including the human, social, environmental and cultural relations connecting the intellectual capital, wealth creation and governance of their regional development. This shows how the networking of the said interrelations and analytical hierarchy framework adopted by the authors to capture the ICT-related developments of smart cities is significant. Significant, if only for the reason that such a framework provides, the authors suggest, the opportunity to capture the triple helix of the urban and regional innovation system of which ICT-related developments are integral parts.

As the chapter points out: since ideas on the creative class first emerged and concepts of the creative industry surfaced with notions of the creative city, an avalanche of urban and regional studies has been undertaken to analyse their cultural and environmental attributes. Despite this, however, an operational conceptualisation of creative infrastructures has as yet not been developed and calls for more profound research. In responding to this call, the authors suggest a city is smart: 'when investments in human and social capital and traditional (transport) and modern (ICT) communication infrastructure fuel sustainable economic growth and a high quality of life, with a wise management of natural resources, through participatory governance'. Furthermore, the chapter suggests cities can become smart if universities and industry support government's investment in the development of such infrastructures.

In moving towards such a representation of smart cities, this chapter's attempt to operationalise the concept of creative infrastructures adopts the triple helix as its point of departure. Here attention focuses on the production of knowledge by universities and industry as an index of intellectual capital tied up in the artefacts of innovations that are patented by industry and licensed in line with the standards laid down by government to regulate such developments.

While offering many critical insights into the political economy of the triple helix, the chapter suggest such studies of knowledge production still reveal little about either the social basis of university, industry and government involvement, or the technical infrastructures of their regional innovation systems. The chapter suggests this is an insight of some significance because universities, industry and government only understand one another when the social and intellectual soil connecting them is fertile enough for the ground they stand on to not only create wealth, but flow through their regulative regime.

The authors of this chapter suggest the inclusion of intellectual capital and wealth creation into such considerations and contours of their technical infrastructures offers the ground for those in university and industry seeking to generate returns from the ICT-related investments that government sets the standards for. Taking such a form, this chapter offers a more profound analysis of the interrelations between the human, social, cultural and environmental relations that not only link intellectual capital to wealth creation and standards regulating the contours of their technical infrastructures, but also connects universities to industry and rates of return from the ICT-related investment which government sets for such electronically enhanced service developments. This networking of the triple helix is then augmented using analytical hierarchy as a framework to cluster and begin measuring the performance of smart cities.

The chapters making up this book are all modified versions of papers which have already been published in scientific journals and leading textbooks and which have been revised to capture, bring together and represent the state of the art on smart cities. It is hoped the body of work collected together under the heading of *Smart Cities: Governing, Modelling and Analysing the Transition* is sufficiently overarching to not only capture the scientific and technical nature of these developments, but the social, environmental and cultural significance of the material also drawn attention to. It is also hoped the manner in which the book goes about capturing this significance is not only sufficiently broad in terms of both scope and ambition, but novel enough for the intellectual capital, wealth creation and standards of regulation that it highlights to motivate others carrying out research into such developments.

Notes

1 http://en.wikipedia.org/wiki/Intelcities

References

Graham, S. and Marvin, S. (2001) *Splintering Urbanism*, Oxon, Routledge.
Hollands, R. (2008) Will the real smart city stand up? *City*, 12: 3, 302–320.
Mitchell, W. (1995) *City of Bits, Space Place and the Infoban*, Cambridge, MA, MIT Press.

Part I
Governing the transition

2 From intelligent to smart cities[1]

Mark Deakin

Introduction

Taking Hollands' (2008) statement on the transition from intelligent to smart cities as its point of departure, this chapter reflects upon the anxieties currently surrounding such developments. In particular, the suggestion that they have more to do with cities meeting the corporate needs of marketing campaigns than the social intelligence required for them to be smart. Focusing on the social intelligence of their environmental and cultural developments, this chapter captures the information-rich and highly communicative qualities of the transition. In particular, the acute social, environmental and cultural challenges that smart communities pose to cities and the critically insightful role that the networks of innovation and creative partnerships set up to embed such intelligence play in the learning, knowledge transfer and capacity-building exercises servicing this community-led transition to smart cities. This, the chapter shall suggest, is what existing representations of smart cities miss, and this account offers as a critical insight on the transition.

Smart city forerunners

Such smart city forerunners as San Diego, San Francisco, Ottawa, Brisbane, Amsterdam, Kyoto and Bangalore are all now setting a trend for others to follow. The other cities keen to follow in their wake and become smart are Southampton, Manchester, Newcastle, Edinburgh, Edmonton, Vancouver and Montreal. It appears the rush to become a smart city has begun to gather apace and, as a consequence, pressure is now growing for cities to become even smarter.

IBM's recent high-profile campaign on smart cities also goes some way to acknowledge this pressure for cities to become smarter. As they state:

> Technological advances [now] allow cities to be "instrumented," facilitating the collection of more data points than ever before, which enables cities to measure and influence more aspects of their operations. Cities are increasingly "interconnected," allowing the free flow of information

from one discrete system to another, which increases the efficiency of the
overall infrastructure ... To [meet] these challenges and provide sustainable prosperity for citizens and businesses, cities must become "smarter"
and use new technologies to transform their systems to optimize the use
of finite resources.[2]

Will the real smart city stand up!

In a recent article, Hollands (2008) asks the question: 'Will the real smart city
stand up?' For according to Hollands, cities all too often claim to be smart,
but do so without either defining what this means, or offering any evidence
to support such proclamations. The all too often 'self-congratulatory' tone
that cities strike when making such claims does not seem to sit well with him.
For while images of the digital city, intelligent city, high-tech district and
neighbourhoods of smart communities abound, in his opinion they all fail
to convey what it means to be smart and why it is important for cities to be
defined in such terms.

In Hollands' (2008) opinion, the validity of any claim to be smart has to be
based on something more than their use of information and communication
technologies (ICTs). Hollands asks this question because cities all over the
world are beginning to do just this and use such technologies as a means of
branding themselves smart. Hollands' anxiety about the 'self-congratulatory'
nature of the claims that cities make to be smart harks back to the image-
building and city marketing campaigns of the 1990s and the competition that
this sparked off between them. Hollands' real fear is that the use of such an
ill-defined notion to spearhead yet another marketing campaign shall end up
merely aggravating the in-built tendency that such marketing campaigns have
to be almost exclusively entrepreneurial in outlook.

Hollands (2008) asks us to be aware that if left to be entrepreneurial,
there is a strong chance that smart cities shall develop in a way that is too
neo-conservative, insufficiently progressive to offer the type of liberating
experience everyone expects from them. For Hollands, the way to avoid the
disappointment of any such neo-conservative route to smart cities lies in
following the clarion cry of those advocating a more neo-liberal path to the
transition. This is because, for him, such a route to smart cities is seen to
be grounded in a critically aware and more realistic understanding of the
transition.

From the intelligent to smart city

In the interests of developing just such a critically aware and realistic
understanding, Hollands (2008) draws particular attention to the work of
Komninos (2002, 2008) on the intelligent city. According to this account of
what it means to be an intelligent city, there are four main components to such
developments, these being:

- the application of a wide range of electronic and digital technologies to communities and cities;
- the use of information technologies to transform life and work within a region;
- the embedding of such ICTs in the city;
- the territorialisation of such practices in a way that bring ICTs and people together, so as to enhance the innovation, learning, knowledge and problem solving that they offer.

This much needed definition of what it means to be an intelligent city is in turn used by Hollands (2008) to clear the way for a vision of cities that are smart because they are:

> … territories with a high capacity for learning and innovation, which is built-in to the creativity of their population, their institutions of knowledge production and their digital infrastructure for communication.

For Hollands (2008: 306) the key elements of this definition relate to the use of networked infrastructures as a means to enable social, environmental and cultural development. While this involves the use of a wide range of infrastructures, including transport, business services, housing and a mix of public and independent services (including leisure and lifestyle services), it is the ICTs of these developments that are of critical importance because they are seen to 'under-gird' (ibid.) all of these networks.

Those ICTs seen as under-girding all of this and lying at the heart of the said networks include mobile and land line phones, satellite TVs, computer networks, electronic commerce and internet services. They are seen as being of critical importance because Hollands (2008) considers the intelligence that such infrastructures embed as the main driving force behind the development of smart cities as regional innovation systems capable of sustaining social, environmental and cultural progress.

Towards smart cities

As Hollands (2008: 315) goes on to state, smart cities, by definition, appear to be 'wired cities', although this cannot be the sole defining criterion because:

> progressive(ly) smart cities must seriously start with people and the human capital side of the equation, rather than blindly believing that IT itself can automatically transform and improve cities.

For Hollands (2008: 316) the critical factor in any successful community, enterprise, or venture, is its people and how they interact. This is because for him, the most important thing about information technology is not the capacity that it has to create smart cities, but the ability that such communications have to be part of a social, environmental and cultural development. That is to say, serve as communications which are smart in the way the deployment of their technologies allow cities to empower and educate people, allowing them to become members of society capable of engaging in a debate about the environment they not only relate to, but cultivate. This, it is stressed, is in turn only made possible when the community of people undergoing such a process of socialisation is able to:

> create a real shift in the balance of power between the use of information technology by business, government, communities and ordinary people who live in cities, as well as seek to balance economic growth with sustainability. ... In a word, the 'real' smart city might use IT to enhance democratic debates about the kind of city it wants to be and what kind of city people want to live in.

To achieve this, Hollands (2008: 316) suggests that cities that really want to be smart shall have to: take much greater risks with technology, devolve power, tackle inequalities and redefine what they mean by smart itself, if they want to retain such a lofty title'.

Some immediate reflections

While Hollands' (2008) image of what it means to be smart tends to start with the worst-nightmare scenario of a city dominated by the entrepreneurial values of the elite few, it is clear this vision of a somewhat unintelligent, neo-conservative and less than liberal representation is soon swept aside by a more progressive alternative. An alternative which, in this instance, uses information technology not to shore up the entrepreneurial values of the city, but to underpin them in a way which is smart. Smart in the sense that in such cities it is the information technologies and not entrepreneurial values which are used as the means to 'under-gird' their social, environmental and cultural qualities.

As a 'best-dream' scenario, this works well to allay any fears that may linger about the purpose of smart cities and ways in which they should be put to work. As with all such visions, however, there are some inconsistencies and omissions in the narrative and storylines this develops as a means to usher in the reworked version of what is being represented. These relate to both the legacy of smart cities and the more contemporary challenges surorunding their development.

In particular, they relate to Hollands' (2008) representation of the 'smart city legacy', which is perhaps just a little too 'fast and furious', in the sense the retrospective he offers relies less on the notion of 'informational cities' advanced by the likes of either Castells (1996) or Graham and Marvin (1996,

2001), and more on Mitchell's (1995, 1999, 2001, 2003) accounts of what it means for the technologies of such infrastructures 'to work smarter not harder'! For while Castells (1996) and Graham and Marvin (1996, 2001) all draw attention to the information technologies of the so-called critical infrastructures (water and drainage, energy and the like), it is Mitchell (1995, 1999, 2001, 2003) that first deployed them in the smart cities laboratory at MIT and sketched out how they make it possible for communities to network the embedded intelligence of smart cities.

The smart card legacy

This can be illustrated by reference to the influence of the smart city laboratory on what Hollands (2008) himself defines as the first smart city, that is, Southampton, the city that first attempted to develop a portal capable of supporting smart card applications. For this initiative, promoted under the triple helix model of university, industry and government, was the first to develop a smart card capable of customising access to a variety of services distributed across the public and independent sectors. It was also the first software development reported as not only capable of supporting the transactional-based logic of multi-application management architectures, but as an enterprise also allowing for services to be added to and removed from the card's dynamic user environment.

The administration of the card scheme involves the processing of personal data in compliance with UK and EU data protection legislation, something that is seen to be of critical importance by the university, industrial and government sectors alike. To comply with this legislation each smart card has a unique identifier, which can be used by all service applications to identify the user; and when transaction information is sent to the data warehouse, this unique identifier is 'one-way' encrypted. This means the unique identifier is scrambled so transaction information cannot be traced back to any user whose personal data is held within the warehouse. However, even though the information held in the data warehouse is stored anonymously, it is still considered to be personal data, due to the fact it is possible to match it with information in other databases.

If service providers wish to share personal data for which they are controller, this must be done for a distinct purpose, underpinned by some formal data sharing protocol. However, where multiple applications are provided by the same data controller, the data collected from these applications can be used in the course of any legitimate interest. This may include cross-matching and trend analysis, where this directly relates to a notified purpose.

It is the ability this portal has to deal with a multiple of transactions simultaneously and as a 'bundles of services' in a real-time environment that has attracted so much attention. This has meant cities shifting attention away from the e-commerce challenges of the enterprise architecture and transaction-based business logic supporting such e-Gov service developments and towards the embedded intelligence of what Halpern (2005) sees as being something really smart.

The informational basis of such communications

The point of this digression into smart cards is simple. It lies in the realisation that it is the legacy of Castells (1996) and Graham and Marvin's (1996, 2001) work undertaken on the informational basis of the communications embedded in such intelligence, rather than the work carried out by Mitchell (1995, 1999, 2001, 2003), that leads us away from the purely technical issues surrounding the business logic of such eGov service developments. That is, away from the technical issues that surround the transaction-based business logic of such developments and towards a more critically insightful examination of the information-rich and highly communicative qualities of the technologies (ICTs) supporting them. In other words, away from the technical aspects of such developments and towards an examination of the social capital, which is not only critical in underpinning the development's informational and communicative qualities, but also equally insightful in revealing the wider environmental and cultural role they play in supporting the transition to smart cities.

Capturing the information-rich and highly communicative qualities of such service developments

The discussion that follows captures the information-rich and highly communicative qualities of these technical, social, wider environmental and cultural developments, the particular methodological challenges they pose and the critically insightful role that the networks of innovation and creative partnerships set up to embed such intelligence also play in the learning, knowledge transfer and capacity-building exercises that service this transition to smart cities. This is what this chapter suggests Hollands' (2008) account of smart cities misses and goes some way to explain why he asks 'the real smart city to please stand up!' For in cutting across the legacy of the transition from the informational to the intelligent, and now smart, city, Hollands' account of the transition is not as well grounded in the informational and communicative qualities of the embedded intelligence that they are built on and stand for.

This, the chapter suggests, is a critical insight of some note, for only in giving such a well-grounded account of the embedded intelligence drawn attention to, is it possible to do what Hollands (2008) asks of the transition. That is 'under-gird' the social capital, which is not only critical in underpinning the informational and communicative qualities of the embedded intelligence that smart cities stand on, but pivotal in gaining a fuller insight into their significance. In particular, in gaining a fuller insight into the wider environmental and cultural significance that their networks of innovation and creative partnerships take on in embedding the intelligence of such an informatics-based and community-led transition to smart cities.

The following shall suggest this insight is equally significant for the reason it takes Hollands' (2008) thinking full-circle by offering an alternative to the top-down entrepreneurial-based business logic that is called for. Something that is realised by turning the top-down entrepreneurial-based business logic on its head and grounding the information-rich and highly communicative qualities of such developments in the community-led logic that is emerging to support the transition. That logic which in real time is aligned with, not against, the cybernetics of the social capital underlying the emergence of smart communities and what their founding networks of innovation and creative partnerships embed as the intelligence of smart cities. Those developments in the cybernetics of social capital whose emergent qualities are now also in the process of being institutionalised in the learning, knowledge transfer and capacity-building exercises which are intelligent in embedding the informatics of this community-led transition to smart cities.

As always, there is another reason for taking an alternative route, and it can be found in the chapter's desire to not only offer an alternative to the entrepreneurial-based business logic embedded in the corporate branding of such developments, but a social and environmental counterpoint to the 'high level' cultural account of smart cities otherwise offered by Cohendet and Simon (2008) in their study of Montreal. For rather than relying on any notion of the 'creative class' offered by Florida (2002, 2004), the account offered here is rooted in an institutional reading of the social capital underpinning the 'environments' that such notions of a 'creative class' attempt to cultivate.

This is because, for the author of this chapter, it is social capital that underpins such environments and that gets bottomed out in this representation of the transition by way of Castells (1996) and Graham and Marvin (1996, 2001), rather than through Mitchell's (1995, 1999, 2001, 2003) investigations into the embedded intelligence of smart cities. The capital that not only underpins such environments, but is also brought to the fore in this chapter's cultural account of the transition. In particular, in the account that surfaces by way of its case-based analysis of the community-led movement in Edinburgh and through an examination of the environments that are being cultivated to support this particular city's transition.

In paving the way for this representation of the transition, the rest of the chapter accounts for what is termed the social capital of smart communities, and for the methodological twist that this particular take on the environmental and cultural attributes of the transition offers. In particular, what this particular take on the transition offers to under-gird the embedded intelligence of smart cities. The embedded intelligence, networks of innovation and creative partnerships that do much to support the environments that Edinburgh cultivates for the reason they go some way to capture what is smart about the city's transition.

This is what is unique about the chapter's account of the transition. For what it manages to do is capture the informational basis of the communications that the transition is built on, rather than rests on as a set of technical claims surrounding the business logic of corporate marketing campaigns. Technical claims which they otherwise rest on and that quickly begin to break down, not because they are weak in themselves, but because they are not grounded in the social capital, environmental and cultural attributes of the very institutions, i.e. communities, that smart cities claim to represent.

The social capital of smart communities

Halpern (2005: 508) speaks of social capital being composed of 'a network; a cluster of norms, rules, values and expectations; and sanctions'. Here communities are understood to form networks and co-operate with one another in accordance with the norms, rules and expectations of their constituents and the power they in turn have to sanction actions taken by fellow members who operate outside the said norms, rules, values and expectations. These in turn are also seen to provide the linkages between members of the community who use these norms, rules and values to bridge divisions that exist in civic society.

Halpern (2005: 509–510) understands ICTs to be forms of social capital and lists several prerequisites for the development of networked communities. These are examined in terms of the potential that networked communities, virtual organisations and managed learning environments have to develop the ecological integrity and equity that allow any such process of democratic renewal to be socially inclusive. As he states:

> While the vast majority of community ICT experiments have to date not met the conditions above [the ecological integrity, equity, democratic renewal, needs and requirements], ICT networks may have great potential to boost local social capital, provided they are geographically "intelligent," that is, are smart enough to connect you directly to your neighbours; are built around natural communities; and facilitate the collection of collective knowledge. They have the potential to connect the work-poor and work-rich.

The methodological twist

The methodological twist in Halpern's (2005) discussions on 'geographically intelligent settlements' lies in the fact that this examination of networks, virtual organisations and managed learning environments precedes those of either the planning or the development of villages and neighbourhoods. This is because for Halpern such networks, virtual organisations and managed learning environments are seen as providing the info-structures needed for communities to collaborate and build consensus on the planning, development and design of high-tech and digitally enabled platforms. That type of planning, development and design which in turn goes a long way to strengthen the norms, rules and values of the said settlements by providing the citizens of these communities with a platform to bridge the gap opening up between the 'work-poor and work-rich'. In particular, bridge the gap that has opened up as a predominately digital divide between the informational basis of the so-called 'work-rich' and 'work-poor', and do this by building a platform capable of tying them together as communities strong enough to carry the environmental weight and cultural expectation (i.e. ecological integrity and equity) of their ongoing regeneration.

Networks, innovation and creative partnerships

So where are these ICT-enabled networks that boost the norms, rules and values of local social capital? Boost them because they are geographically 'intelligent', that is, smarter, because they are sufficiently innovative to connect 'villagers' directly to 'neighbours' and do this by virtue of being based on creative partnerships which are built around 'natural' communities. Around natural communities that facilitate the generation of collective knowledge that they can draw upon in order to meet the expectations of their on-going generation? Contrary to popular belief, such urban regeneration programmes are not limited to the UK, but can be found throughout Europe. For examples can be found in Edinburgh, Helsinki, Glasgow and Dublin (Deakin et al., 2005; Deakin and Allwinkle, 2006).

The following offers a very brief account of the networks, innovation and creative partnerships underlying the villages and neighbourhoods of the geographically intelligent and smart community regeneration programmes underway in Edinburgh as part of the city's Social Inclusion Partnership (SIP). This account will examine the city's smart regeneration programmes for the Wester Hailes and Craigmillar communities (Deakin and Allwinkle, 2007). An account of the urban regeneration partnerships governing the planning, development and design of Wester Hailes and Craigmillar can be found in Hastings (1996) and Carley et al.'s (2000) report on such developments (see also Carley and Kirk, 1998). The development of these networks and innovations are also reported on by Slack (2000), Malina (2001, 2002) Malina and MacIntosh (2004) and McWilliams et al. (2004). Malina and Ball (2005) have also reported on the development of social capital, and the environmental and cultural attributes emerging from these innovations. They have also addressed the question of whether the villages and neighbourhoods emerging are not just geographically 'intelligent', but 'smart-er' in the sense that their communities connect 'villagers' to their 'neighbours'.

The following shall cut across these reports and draw attention to the creative partnership(s) emerging from the urban planning, development and design of the villages and neighbourhoods making up the communities in question. Similar developments under way in Helsinki are reported on by

Sotarauta (2001) Kostiainen and Sotarauta (2003) and Sotarauta and Srinivas (2006). Dabinett (2005) also reports on the situation emerging in Glasgow and Dublin.

This also goes some considerable way to turn around what Hollands (2008) suggests is the 'should' of smart cities by offering an examination of what they actually do. In short, it serves to show that by 'taking risks with technologies, devolving power and tackling inequalities', it is possible to turn things around and hold cities up as a clear demonstration of 'what it means to be smart'. In other words, provide the material to match the means with the ends.

myEdinburgh.org

As an ICT-enabled network, myEdinburgh.org is innovative because it gets beyond the user-centred environment of the smart card legacy by providing an information portal and community grid for learning (CGfL). For this particular information portal provides citizens with the user-friendly tools for communities to access learning opportunities. Within this environment, the CGfL provides the infrastructure needed for citizens to learn about the planning, development and design of their city and engage in local decisions made about the promotion of urban villages and neighbourhoods as sustainable communities under the city's urban regeneration strategy.

The Edinburgh Learning Partnership, composed of representatives from local government agencies, the education sector, voluntary groups and private-sector businesses, provides the creative basis for the networking and innovation that the said portal and grid offers access to. As a city-wide venture, the collaboration seeks to encourage and facilitate initiatives that are creative in widening access to and increasing participation in learning activities, particularly those that target the disadvantaged. The key aims of the partnership can be summarised as follows:

*   to provide citizens with ICT 'taster' sessions in local, accessible venues; specifically targeting citizens identified as 'digitally excluded' (for example, citizens living in Edinburgh's SIPs);
*   to support community and voluntary organisations in the procurement, use and development of ICTs, including training staff to access and maintain the information portal;
*   to develop a Community Grid for Learning (CGfL);
*   to use the said grid for learning to build capacity and engage citizens in local decision making;
*   to transfer the knowledge required to participate in the planning, development, design and layout of villages and neighbourhoods and democratic renewal needed for this process of modernisation to govern their development of self-sustaining communities.

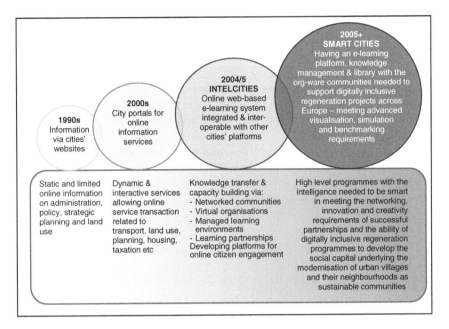

Figure 2.1 The development of digitally inclusive regeneration programmes
Source: Deakin and Allwinkle (2007)

E-learning, knowledge transfer and capacity building

The resulting e-learning platform makes it possible for the online service applications being demonstrated to be integrated with the knowledge transfer and capacity-building technologies needed to meet the interoperability requirements of such developments.

This allows the citizens, communities and organisations in question to collaborate and build consensus on the competencies, skills and training required for the development of these online services to support urban regeneration programmes. Together, these networks of innovation and creativity underpin the partnerships responsible for embedding the intelligence of those technologies, skills and training exercises, making it possible for citizens to engage in this process and show how active participation in the said venture is smart. That is to say, not only intelligent because it draws upon such services, but smart for the reason that their participation in this venture develops the social capital that sets the normative standards and rule-based logic of the very civic values governing the ecological integrity and equity of their environmental and cultural attributes. The environmental and cultural attributes are in this instance responsible for promoting the deeply rooted process of democratic renewal which their very participation in this venture is key to unlocking.

This is made possible because:

- the ICT-enabled networks are innovative in developing an e-learning platform based on open-source technologies that are interoperable across online services;
- satisfying the need for a formal learning community, this high-tech, digitally enabled network allows for the planning, development and design of the online services that are needed for regeneration programmes to support SIPs;
- these partnerships, in turn, allow the ecological integrity, equity, democratic norms, rules and values of the applications being demonstrated to be integrated with the e-learning, knowledge transfer and capacity-building technologies supporting the regeneration programmes;
- the citizens and communities can then collaborate and build consensus on the competencies, skills and training needed for the development of online services required to support the quintessentially civic values of this regeneration programme;
- together the said networks, innovations and partnerships create the trust needed to engage citizens and show how the active participation of communities in digitally inclusive decision making is both intelligent in developing the social capital – norms, rules and civic values – of the ecological integrity and equity underlying the modernisation and smart enough to support it;
- here the ecological integrity and equity of the democratic renewal take the form of consultations and deliberations in government- and citizen-led decision making which engages citizens as members of a community participating in this modernisation;
- the resulting platform supports the distribution, storage, retrieval of learning material, skill packages and training materials needed for such engagement and participation to bridge the digital divides that currently exist, build the capacity that exists for inclusive decision making and transfer the knowledge required for citizens to bond with one another as members of a community. The ecological integrity and equity of this socially cohesive process takes the form of decisions over the development's ecological footprint, biodiversity, environmental loading and cultural identity.

The democratic renewal does this by promoting the shift from government- to citizen-led decision making within the community. This involves the use of advisory groups, discussion boards, opinion polls, focus groups, petitions, citizen juries, ballots and online voting. These are all part of the visioning and scenario-building exercises for gaining consensus on the norms of energy consumption, waste and emissions, as a set of rules underlying the social equity, environmental justice and cultural heritage of the democratic renewal governing this process of modernisation.

As the standards of knowledge transfer and capacity building are drawn from a review of leading city portals across Europe and from the benchmarking of their respective CGfLs against stakeholder requirements, the intelligence embedded in these applications should be smart enough for the physiology of the built environment that cities (re)generate and which the urban planning, development and design of villages and neigbourhoods in turn cultivate, to not be overloaded, but able to carry the weight of their transformation into self-sustaining communities.

All this is in contrast to how villages and neighbourhoods are traditionally represented as self-sustaining communities and offers an illustration of how the intelligence being embedded in this particular smart city can stand up and be counted for the progress it is making. In particular, in terms of the progress this smart city is making to realise the opportunities the transition offers not to fall back on unsubstantiated claims as to what is smart, but for the city to move forward on a progressive and well-founded social, environmental and cultural agenda.

Conclusions

Taking Hollands' (2008) statement about the 'unspoken assumption' as the point of departure, this chapter has reflected upon the anxieties currently surrounding the transition from intelligent to smart cities. In particular, the suggestion that such developments have more to do with cities meeting the corporate branding needs of marketing campaigns than the social intelligence required for them to be smart.

Working on the assumption that any attempt to overcome such anxieties means cities shifting attention away from the needs of the market and towards the intelligence required for them to be smart, this chapter begins to set out a less presumptious and more critically aware understanding of the transition. In particular, an understanding based on the legacy of Castells (1996) and Graham and Marvin's (1996, 2001) research carried out on the informational basis of the communications embedded in such intelligence, rather than offered by Mitchell (1995, 1999, 2001, 2003). An understanding that in turn leads away from the purely technical issues that surround the business logic of eGov-related service developments. That is, away from the technical issues that surround the business logic of such developments and towards a more critically insightful appreciation of the information-rich and highly communicative qualities of the technologies supporting them. Away, in that sense, from the technical aspects of such developments and towards an examination of the social capital that is not only critical in underpinning their informational and communicative qualities, but also insightful in revealing the wider environmental and cultural role they play in supporting the transition to smart cities.

Armed with this critical insight, the chapter has sought to capture the highly communicative qualities of such service developments, the particular methodological challenges that smart communities pose cities and the critically insightful role that the networks of innovation and creative partnerships set up to embed such intelligence, play in the learning, knowledge transfer and capacity-building exercises servicing this community-led transition to smart cities.

This is what Hollands' (2008) account of smart cities misses and goes some way to explain why he asks 'the real smart city to please stand up'. For in cutting across the social capital and environmental capacities of the transition from intelligent to smart cities, the representation that surfaces is insufficiently grounded in the informational and communicative qualities of the embedded intelligence that not only underpins such developments, but also offers the means to do what Hollands asks of them. That is, 'under-gird the social capital that is not only critical in underpinning their informational and communicative qualities, but also insightful in revealing the wider environmental and cultural role that their networks of innovation and creative partnerships play in supporting the development of a community-led transition to smart cities.

This insight is equally significant for the reason that it takes Hollands' (2008) thinking full-circle and in doing so offers the very alternative to the top-down entrepreneurial-based that business logic called loud and hard for. This is achieved by grounding the developments under review in what are referred to as the community-led logic of the transition. The logic that in real time is aligned with the cybernetics of the social capital underlying the emergence of smart communities and what their networks of innovation and creative partnerships embed as the intelligence of such developments. Those developments whose embedding has now become institutionalised in the learning, knowledge transfer and capacity-building exercises supporting the environments that this very community-led transition to smart cities does much to pave the way for.

By focusing on the social capital of collaborative platforms and consensus building, it has also become possible to recognise the critical role that networks, innovation and creative partnerships play in representing places that are not only sites of ecological integrity, social equity, environmental justice, cultural heritage and democratic renewal, but locations where socially inclusive decision making can institutionalise the civic values that are required for the regeneration of urban villages and the neighbourhood as self-sustaining communities.

Having reported on an ICT-enabled network that provides just such an information portal and CGfL, the chapter has reviewed the innovative features of partnerships which are creative in organising the development of the necessary technologies, skills and training exercises. The partnerships found to be successful are those making it possible for regeneration programmes to engage citizens and demonstrate how the embedded intelligence of community participation is smart. The evidence generated from the Edinburgh case study is valuable because its findings reveal that successful partnerships not only develop the norms, rules and civic values of any regeneration, but also express the programme's ecological integrity, social equity, environmental justice and cultural heritage as part of an underlying venture in democratic renewal.

The main findings of this case study can be summarised as follows:

- Questions about the critical role of networking, innovation and the creativity of partnerships have previously remained unanswered because many believe they are resources that can easily be assembled and that can be left to develop as virtuous circles of mutually reinforcing actions.
- This underestimates the extent of the embedded intelligence, networks, innovation and creativity needed to build partnerships and be successful in meeting their capacity building and knowledge-transfer requirements.
- Many such partnerships tend to represent little more than short-term measures at self-help and exercises in community learning on matters related to how best to pull yourself up by your own bootstraps. That so-called self-help aesthetic stands in opposition to a collective knowledge of the social, environmental, cultural and civic values underpinning this process of democratic renewal and those supporting any such transition to smart cities.

The chapter has also found that it is the e-learning platform emerging from these developments that makes it possible for the online service applications being demonstrated to integrate with the knowledge transfer and capacity building needed to meet the interoperability requirements of such developments. This allows the citizens, communities and organisations in question to collaborate and build consensus on the competencies, skills and training needed for the development of the online services required to support urban regeneration programmes. Together, the networks of innovation and creative partnerships responsible for organising the development of these technologies, their requisite skills and training exercises make it possible to engage citizens and show how their active participation is valuable; in particular, to show how their active participation is valuable because it embeds the intelligence of smart cities, not as some entrepreneurial-based business logic, capable of being corporately branded, but as the social capital (norms, rules and civic values) that communities are smart enough to draw upon as a means of governing the ecological integrity, social equity, environmental justice and cultural heritage of the urban regeneration programmes they are subject to.

This, the chapter suggests, is being achieved by promoting the shift from government- to citizen-led decision making within the community. It involves the use of advisory panels, discussion boards, opinion polls, focus groups, petitions, citizen juries, ballots and online voting. Panels, boards, polls, groups and juries, all forming part of the visioning and scenario-building exercises set up to gain consensus on the norms of energy consumption, waste and emissions, as a set of rules underlying the equity, justice and heritage of the democratic renewal governing the modernisation that smart cities are tied up with.

Notes

1 This is a modified version of an article by Deakin, M. and Al Wear, H., From Intelligent to Smart Cities, *International Journal of Intelligent Buildings*, 2011, 3 (3): 140–152.
2 IBM's website on smart cities: www-935.ibm.com/services/us/gbs/bus/html/smarter-cities.html

References

Carley, M. (1995) The Bigger-Picture: Organizing for Sustainable Urban Regeneration, *Town and Country Planning*, 64 (9): 236–239.
Carley, M. and Kirk, K. (1998) *Sustainable by 2020?: A Strategic Approach to Urban Regeneration for Britain's Cities*, Bristol, Policy Press.
Carley, M. Chapman, A. Hastings, K., Kirk, and Young, R. (2000) *Urban Regeneration Through Partnership: A Study in Nine Urban Regions in England, Scotland and Wales*, Bristol, Policy Press.
Castells, M. (1996) *Rise of the Network Society: The Information Age*, Cambridge, Blackwell.
Cohendet, P. and Simon, L. (2008) Knowledge-intensive Firms, Communities and Creative Cities, inn Amin, A. and Roberts, J., eds, *Community, Economic Creativity and Organisation*, Oxford, Oxford University Press.
Curwell, S. and Lombardi, P. (2005) Analysis of the IntelCities Scenarios for the City of the Future, inn Miller, D., ed., *Beyond Benefit Cost Analysis*, Andover, Avebury.
Curwell, S., Deakin, M., Cooper, I., Paskaleva-Shapira, K., Ravetz, J. and Babicki, D. (2005) Citizens' Expectations of Information Cities: Implications for Urban Planning and Design, *Building Research and Information* 22 (1): 55–66.
Dabinett, G. (2005) Competing in the Information Age: Urban Regeneration and Economic Development Practices, *Journal of Urban Technology* 12 (3): 19–38.
Deakin, M. (2009a) The IntelCities Community of Practice: The eGov Services Model for Socially-Inclusive and Participatory Urban Regeneration Programmes, in Reddick, C., ed., *A Handbook of Research on Strategies for Local e-Government Adoption and Implementation: Comparative Studies*, Hershey, IGI Global.
Deakin, M. (2009b) Towards a Community-based Approach to Sustainable Urban Regeneration, *Journal of Urban Technology*, 16 (1):
Deakin, M. and Allwinkle, S. (2006) The IntelCities Community of Practice: The e-Learning Platform, Knowledge Management Systems and Digital Library for Semantically-Interoperable e-Governance Services, *International Journal of Knowledge, Culture and Change in Organizations*, 6 (2): 155–162.
Deakin, M. and Allwinkle, S. (2007) Urban Regeneration and Sustainable Communities: The Role of Networks, Innovation and Creativity in Building Successful Partnerships, *Journal of Urban Technology*, 14 (1): 77–91.
Deakin, M., Allwinkle, S., Campbell, F. and Van Isacker, K. (2005) The IntelCities e-Learning Platform, Knowledge Management System and Digital Library, in Cunningham, M. and Cunningham, P., eds, *Innovation and the Knowledge Economy: Issues, Applications, Case Studies*, Amsterdam, IOS Press.
Florida, R. (2002) *The Rise of the Creative Class: And How It's Transforming Work, Leisure, Community and Everyday Life*, New York, Basic Books.
Florida, R. (2004) *Cities and the Creative Class*, London, Routledge.

Graham, S. and Marvin, S. (1996) *Telecommunications and the City*, London, Routledge.

Graham, S. and Marvin, S. (2001) *Splintering Urbanism*, London, Routledge.

Halpern, D. (2005) *Social Capital*, Bristol, Policy Press.

Hastings, A. (1996) Unravelling the Process of Partnership in Urban Regeneration Policy, *Urban Studies*, 33 (2): 253–268.

Hazlett, S. A. and Hill, F. (2003) E-government: The Realities of Using IT to Transform the Public Sector, *Managing Service Quality*, 13 (6): 445–452.

Hollands, R. (2008) Will the Real Smart City Stand Up? *City*, 12 (3): 302–320.

Holm, J. and Wambui-Kamara, M. (1997) The Participatory and Consensus-Seeking Approach of Denmark, in Lafferty, W., ed., *Sustainable Communities in Europe*, Andover, Avebury.

Janssen, M., Kuk, G. and Wagenaar, R. W. (2008) A Survey of Web-based Business Models for e-Government in the Netherlands, *Government Information Quarterly*, 25 (2): 202–220.

Komninos, N. (2002) *Intelligent Cities: Innovation, Knowledge Systems and Digital Spaces*, London, Spon Press.

Komninos, N. (2008) *Intelligent Cities and Globalisation of Innovation Networks*, London, Taylor & Francis.

Kostiainen, J., and Sotarauta, M. (2003) Great Leap or Long March to Knowledge Economy: Institutions, Actors, and Resources in the Development of Tampere, Finland, *European Planning Studies* 10 (5): 415–438.

Landry, C. (2008) *The Creative City*, London, Earthscan.

McWilliams, M., Johnstone, C. and Mooney, G. (2004) Urban Policy in the New Scotland: The Role of Social Inclusion Partnerships, *Space and Polity* 8 (3): 309–319.

Malina, A. (2001) Electronic Community Networks, *Journal of Community Work and Development*, 1 (2): 67–83.

Malina, A. (2002) Community Networks and Perception of Civic Value, *Communications*, 27: 211–234.

Malina, A. and Ball, I. (2005) ICTs and Community: Some Suggestions for Further Research in Scotland, *Journal of Community Informatics*, 1 (3): 66–83.

Malina, A. and MacIntosh, A. (2004) Bridging the Digital Divide: The Development in Scotland, in Ari-veikko-Anttroiko et al., eds, *eTransformation in Governance*, Hershey, PA, Idea Group Publishing.

Mitchell, W. (1995) *City of Bits: Space, Place, and the Infobahn*, Cambridge, Massachusetts, MIT Press.

Mitchell, W. (1999) *E-Topia: Urban Life, Jim But Not as You Know It*, Cambridge, Massachusetts, MIT Press.

Mitchell, W. (2001) Equitable Access to an On-line World, in Schon, D., Sanyal, B. and Mitchell, W., *High Technology and Low-Income Communities*, Cambridge, Massachusetts, MIT Press.

Mitchell, W. (2003) *Me ++: The Cyborg-self and the Networked City*, Cambridge, Massachusetts, MIT Press.

Rallet, A. and Torre, A., eds (1995) *Economie industrielle et économie spatiale*, Paris, Economica.

Scholl, H. J. (2006) Electronic Government: Information Management Capacity, Organizational Capabilities, and the Sourcing Mix, *Government Information Quarterly*, 23: 73–96.

Slack, S. (1999) *The Problematics of Community Information Services: The Case of CCIS*, Edinburgh, Edinburgh University.

Slack, S. (2000) The Dialectics of Place and Space: On Community in the Information Age, *New Media and Society*, 2 (3): 313–334.

Sotarauta, M. (2001) Network Management and Information Systems in Promotion of Urban Economic Development: Some Reflections from CityWeb of Tampere, *European Planning Studies*, 6: 693–706.

Sotarauta, M. (2005) Tales of Resilience from Two Finnish Cities: Self-Renewal Capacity at the Heart of Strategic Adaptation, in Duke, C., Osborne, M. and Wilson, B., eds, *Rebalancing the Social and Economic: Learning, Partnership and Place*, Leicester, NIACE.

Sotarauta, M. and Srinivas, S. (2006) Co-Evolutionary Policy Processes: Understanding Innovative Economies and Future Resilience, *Futures*, 38 (3): 312–336.

3 E-governance as an enabler of the smart city[1]

Krassimira Paskaleva

Introduction

At the onset of the twenty-first century, the challenge to urban governments in the countries of the European Union (EU) is to use information and communication technologies (ICT) to develop their cities competitively in the increasingly complex and interconnected world. This, however, requires not just innovation in technology, but also institutional reforms and policies that engage citizens in democratic activities to improve urban competitive advantages and local prosperity. E-governance, generally referred to as rules, processes and behaviour that affect the way in which powers are exercised at different levels, particularly as regards openness, participation, accountability, effectiveness and coherence (the 'good governance' paradigm), has emerged as an important driver of the necessary transformation (European Commission (EC), 2007). At the same time, strategic EU development priorities re-enforce the principles of 'sustainability' and 'competitiveness', focusing on strengthening the durability of the local potentials and capacities in the global arena of highly competitive markets and localities (EC, 2002).

At the beginning of the new millennium, throughout Europe, ICT-relevant policies are becoming more and more significant. But the urban applications of these policies are yet to evolve as a means to guide local e-governance strategically and competitively. In response to this demand, the recent IntelCities Project[2] (www.intelcitiesproject.com) set up a framework of 'city e-governance' based on the principles of good governance. This offers a coherent interdisciplinary conceptual framework for public policy and practice, which links e-government with e-democracy to involve citizens in local decision-making and bridge the academic–practitioner–policy divide.

In the same light, recent literature surrounding the impacts of ICT suggests that it can be a powerful tool for building the collaborative digital environment that enhances the intelligence capacity of localities and that helps them become a 'smart city', a concept which is intrinsically linked not only to that of the 'knowledge-based economy', where innovation and technology are the main drivers of regional and local growth (Torres et al., 2006), but also to the notion of 'collective community intelligence', which underlines such capacities and contributes to their success (Baron et al., 2000). In the context of the present study, the smart city is defined as one that takes advantage of the opportunities offered by ICT to increase local prosperity and competitiveness – an approach which implies integrated urban development involving multi-actor, multi-sector and multi-level perspectives. In it, arguably, the notion of e-governance provides a strategic dimension to the interface between local government and ICT, the urban community and the stakeholders in promoting the digital environment and collective intelligence necessary to fulfil their democratic mandates in the political, socio-economic and strategic development urban agenda (Odendaal, 2003; Steins, 2002).

This chapter presents empirical evidence for the nature of the e-governance progress in cities across Europe, which reveals that the focus of the e-city governance concentrates on the delivery of services and information, rather than on areas that enhance the overall competitiveness of the community through digital collaboration and collective intelligence. Amidst the growing evidence of city e-governance acting as an enabler of the 'smart city', the study suggests that e-governance is yet to evolve as an effective medium facilitating urban competitiveness. The present work adds a new dimension to some earlier research by the author of this chapter which defined an e-readiness index to assess progress of e-governance in the EU states (Paskaleva-Shapira, 2007, 2008). Here, the emphasis is on the key dimensions that are embedded in the development of the smart city. In fact we elaborate a conceptual approach for examining the interface between city e-governance and the smart city. We do this by exploring current trends in city e-governance in Europe and by focusing on particular dimensions relevant to the notion of the smart city: networks and partnership development, integrated e-services, e-participation and policy innovation.

The first section of the chapter describes the conceptual model of city e-governance, the interface with the smart city notion and the study approach to assessing readiness of city e-governance. The second section presents the results of a comprehensive survey study on the progress of city e-governance in Europe in regard to the smart city. The third section offers propositions on the future of city e- governance and the implications for strategic policy innovations to foster smart cities. The concluding remarks provide reference to the strategic agenda of city e-governance in Europe and offer a model for the enduring smart city.

Research background

In the new era, digital networks increasingly take hold and reshape the way people live, communicate and work. At a time of great challenges for the world's urban communities, the question of the role of governance – the sum of the many ways individuals and institutions, public and private, plan and manage the common affairs of the city (UN-HABITAT, 2002) – is increasingly raised, suggesting the need for a more community-based model of e-governance, with ICT providing greater connectivity and engagement between the players and the stakeholders in a dynamic and increasingly virtual environment.

What transpires from the literature surrounding e-governance and smart cities is the importance of the role of governance in communication networks (Preissl and Muller, 2006), the reshaping of government's relationship with citizens and the social learning and knowledge transfer challenges, among others (Torres et al., 2005). Current literature on ICT and smart communities strongly suggests that the use of ICT in local government can sufficiently enhance the management and functioning of cities. For example, Coe, Paquet and Roy (2001) found that the application of ICT locally leads to economic, social and political transformations encapsulated by the new smart community movement. In the countries of the European Union, these changes have been driven by long-term strategies of sustainability and competitiveness aimed at building competitive cities through creative knowledge networks, strategic planning, integrated urban management and the successful exploitation of ICT for the desired urban pursuits of sustainability and prosperity (Cooper et al., 2005). This means that the adoption of ICT in the urban community needs to explicitly relate to its specific agenda of competitiveness, calling for governance and social cohesion (Buck et al., 2004).

But in responding to the demand for new knowledge, it is necessary to explore more fully how ready cities in Europe are for the challenges of the smart city. This chapter adds new dimensions to the existing terrain. Of particular interest is the phenomenon of e-governance, with the attention focusing on the networking opportunities that e-governance can facilitate between city governments and the relevant stakeholders, the increasing use of integrated e-services which reflect urban development issues that impact on the competitiveness of the city, and citizens' e-participation, which provides opportunities for engaging in democratic activities and policy-making processes virtually, and finally, the role of policy innovation for driving the change. The current section attempts to describe our model of city e-governance and its relevance to the smart city.

City e-governance

Most recent trends in e-government, viewed traditionally as the provision of services by government through digital means (Moon, 2002), show a dramatic shift from information diffusion towards community-based interactive models of services, participation and innovation. In parallel, the construct of e-governance has emerged as a novel platform for governing in twenty-first-century Europe. As a result, the success of e-governance of nations, regions, localities and organisations is being increasingly assessed by an 'e-readiness' index, viewed as the degree to which a community is prepared to participate in the networked world (Global e-Policy eGovernment Institute, 2003).

This systematic approach to measuring the success of local communities in the digital age assumes a more integrated concept of city e-governance which reflects the perspectives of the stakeholders – government, citizens and businesses (OECD, 2003; Zimmermann, 2005). Accordingly, recent advances in research and policy have affirmed e-governance as the use of digital technologies by government agencies to facilitate effective decision-making and improve public policies in the local communities by transforming relations with citizens, businesses and other arms of government. Implementation should seek broader benefits for the city and its people – better service to citizens, more effective government, enhanced local democracy and improved decision-making processes. A detailed stakeholder requirements analysis undertaken by the IntelCities Consortium between 2004 and 2007 has consolidated the work in this area as the basis of developing a city e-governance framework (see also Castells and Hall, 1994; Snellen and Van Der Donk, 1998; Van Den Berg and Van Winden, 2002) which is regarded as representative of the specifics of all user groups and relevant to a range of strategic policy agendas, such as intelligent communities, urban competitiveness and sustainability, participatory democracy, learning communities and system innovation. Thus, the framework links e-government, which the EU regards as a means of achieving world-class public administration and of providing a major economic boost and more democratic development in Europe (EC, 2007), both by way of e-democracy which calls for more public participation in governing and through community development steering, where the political will needs to be delivered based on new relationships between government and the citizens (Fifth Worldwide Forum on e-Democracy, 2004). It also reflects the goals of new public management, with reformers calling for public management becoming a vibrant field of partnering, participation and networking (Barzelay, 2001). In particular, the IntelCities city e-governance model establishes four main relational schemes which are interrelated:

- links and interactions within government;
- links and interactions between government and businesses/citizens;
- links and interactions within and between NGOs;
- links and interactions within and between communities.

In its wholeness, it also addresses the technical integration, socio-economic, legal and e-inclusion issues of implementation and identifies ways to innovation, particularly with regard to the integration of front- and back-offices and government modernisation. It also defines the strategies in key e-governance policies, as well as themes central to implementation in relevance to building and developing the technical and knowledge infrastructure that is able to link together city e-government with strategic development and participatory e-governance processes. In essence, the city e-governance policy framework is

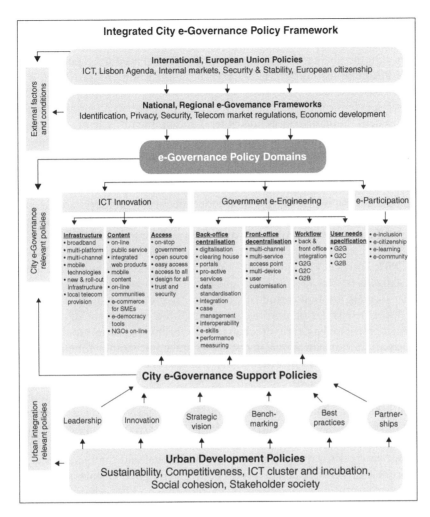

Figure 3.1 Integrated city e-governance model

Source: Paskaleva-Shapira, 2007. © *Copyright 2007 IEEE – All rights reserved.*

an e-inclusive and citizens-centred system innovation model which is relevant to other urban development policies, as illustrated in Figure 3.1.

Three interrelated issues arise from our model of city e-governance. Firstly, that e-governance relates to the relationship between individuals, interest groups, institutions and service providers in the ongoing business of government, which implies networking, collaborative environments and integrated products and activities. Secondly, that e-governance includes civil society as a key player in government processes (Rakodi, 2001), which implies citizens' participation and engagement. And thirdly, that governance is a dynamic phenomenon that is influenced by political circumstances as well as by the relationships between the various actors that contribute to the aforementioned dynamic. The role of local government is therefore interesting in this regard. On the one hand, local government is best placed to respond to local needs meaningfully, appropriately and in a timely way, by providing user-relevant and user-centred services developed as integrated deployment of virtual services (education, administration, transport, urban planning, property development, human services and others) and on the other hand, because of their democratic mandate and leading role in the urban community, they are best placed to promote city-wide collaborative environments to integrate with the relevant stakeholders and the residents that would use or benefit from such services.

Recent developments suggest that under the e-governance paradigm some public managers start to shift away from emphasising producer concerns such as cost efficiency, to focus on user satisfaction and control, flexibility in service delivery and network management with internal and external parties (Tat-Kei Ho, 2002; Torres et al., 2005). But studies also suggest that building relations with citizens is slow (Noordhoek and Saner, 2004) and the perception that public services are failing and of poor quality grow (Oakley, 2002). So the search for new styles of governance through the engagement of citizens is increasingly viewed as a way to change such feeling and of improving citizen participation in urban governing so as to make governance function better than it currently does. The interactivity provided is expected to empower citizens to actively participate in decision-making. Clearly, e-governance acquires an increasingly developmental role for local government which stretches beyond the traditional role of service provision (Frissen, 1997), extending to overall social and economic urban development, thus providing government and citizens with the challenge of being more proactive and inventive.

What emerges from our city e-governance model and the literature is the importance of network collaboration between city governments, municipalities, other institutions, public–private partnerships and the urban stakeholder groups. Moreover, establishing networks to enable the implementation of an e-governance strategy in all its guises not only requires institutional integration, but is key to success. The integrated services that can come from public–private partnerships, from networks with other state agencies and establishing opportunities for integrated one-stop services from the city portals, clearly marks the need to collaborate with the private sector as well in implementing urban IT strategy to enable collaborative networks around e-governance. Associated with this is the emphasis on consultation and participation with communities, allowing for continuous two-way communication between city governments and their constituents for greater transparency of decision- and policy-making. A move towards 'joined-up' local government, in terms of integration across the department and other government agencies, is also necessary. Overall, a more strategic focus for local governance that manages change and embraces innovation seems a must. And policy priorities underpin all of these processes (Borja and Castells, 1996). In summary, networks and part-nerships, integrated e-services and citizens' participation will most certainly determine the role of e-governance in building the smart city. Innovation of policy would arguably have a pivotal impact too. Together, these factors pro-vide the framework for analysis of city e-governance in the EU.

The smart city and the interface with e-governance

In the knowledge-based economy driven by technological change and innovation, new challenges are emerging. Two sets of forces tend to increase the interest in smart communities and e-governance: the reinforced importance of city competitiveness and the new potentialities of governance made possible by ICT. Yet, to date, there are still only a few studies that explore the relationships between e-governance and the smart city.

Internationally, there is a vast body of literature that has informed much of the recent thinking on smart communities. As early as 1994 Saxenian described the key determinants of sub-national competitiveness, ascribing the success of Silicon Valley to the establishment of network systems that compete intensely while also collaborating in both formal and informal ways with each other and with local institutions such as universities or public agencies. It has been argued since that network systems, by linking public, private and academic organisations, facilitates collective learning and the intelligence that provides a competitive advantage to the locality (Horan and Wells, 2005). Therefore, the success of the locality is said to be determined, in large part, by its effec-tiveness in gathering and using the entire stock of knowledge and technology of the community through sustained knowledge transfer – intensive relations, networks of individuals, firms and institutions linked electronically.

So the notion of smart community refers to the locus in which such net-worked intelligence is embedded. The result is a society composed of more network-based governance patterns. But although the future of electronic networks may be at the heart of the smart communities movement, gover-nance is dependent on more complex sets of social ties. It requires durable and dynamic collaborative relations between sectors in local systems of gov-ernance to generate new knowledge that can be translated into products and services (Leadbeater, 1999). Thus a smart city can be defined as an urban com-munity within which citizens, organisations and governing institutions deploy ICT to transform their locality in a significant and fundamental way (Eger, 1997). In the information age, smart cities are intended to promote economic affluence and quality of life within the community. In the context of today's globalisation, this means competitiveness, which for cities translates into:

- the ability and capacity of a city to provide a vibrant urban economy and society;
- an overall city attractiveness which benefits all consumers – businesses, citizens and visitors;

- the effective organisation and governance of local innovation systems which enhance the management of urban resources and improve both situational conditions and market performance (see, for example, Paskaleva-Shapira, 2008).

Although here the emphasis is very much on the dimensions of competitiveness in the global knowledge-based economy, with smart cities, the linking together of government, business and citizens, is said to provide an opportunity for enhancing citizens' participation and influence in local decision-making. Leadership is also required for political and social innovation to strengthen local capacities and advantages by innovative e-services (Coe et al., 2001). So it appears worthwhile exploring how well city e-governance is placed to enable the emergence of the smart city.

City e-governance readiness

In previous research on e-governance, we developed the 'city e-governance readiness' index which, in its wholeness, reflects the ability of the overall society to benefit from ICT cohesively and strategically (Paskaleva-Shapira, 2007, 2008). Figure 3.2 represents our vision of understanding readiness in city e-governance.

Based on this vision of city e-governance, an assessment approach was proposed that can measure the e-governance-ready city. We distinguish *five key dimensions* of *readiness* for city e-governance and their main categories: 1) *the e-governance integrated framework dimension*, focusing on the strategic vision, concept and definition, legislative framework, learning and diffusion of 'best practices', stakeholder participation, integration, implementation, monitoring, evaluation, review and innovation; 2) *the e-service dimension*, which includes all types of e-services that meet the needs and requirements of citizens and businesses making information available and providing interactive public services online and accessible for all using multi-platform access to create connections and interaction between citizens, enterprises and administrations and the IT systems that support their delivery based on broadband connection, interoperability frameworks, open standards and secure and reliable information infrastructure; 3) *the public management dimension*, which includes developing instruments of data management and knowledge management and which incorporates good governance practices in the operations of the local government, back-office centralisation, front-office decentralisation, back-front office integration, human development, networking and so forth; 4) *the citizens' participation dimension*, focusing on the continuous interaction between government and the citizens in the making of urban governing and accounting for key attributes such as users, tools, services, functions, usage, inclusion and empowerment; and 5) *policy innovation* as the fifth key dimension, dealing with the continuous review and enhancement of policies that are directly relevant to e-governance (e-services, government modernisation,

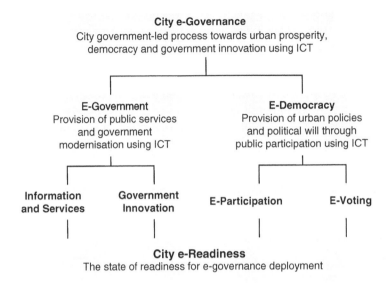

Figure 3.2 Model for city e-governance readiness
Source: Paskaleva-Shapira, 2007. © *Copyright 2007 IEEE – All rights reserved.*

e-participation), urban integration policies (social cohesion, inclusive society, curb of digital divide, stakeholder collaboration, participation in decision making, liveable community), the competitive city (sustainable economy, society and environment) and other relevant regional, national, EU and international policies. In the next section, we show how these dimensions interact in some European cities through a selected set of qualitative indicators and link these developments to the future of the smart city.

City e-governance progress in Europe

Study methodology

The results presented below stem from two main sources:

a. Electronic survey carried out in 2005 in 12 EU cities – Athens (Greece), Reykjavik (Iceland), Venice (Italy), Linz (Austria), Karlsruhe (Germany), Barcelona (Spain), Siena (Italy), Marseille (France), Nice (France), Manchester (United Kingdom), Florence (Italy) and Camden (London, United Kingdom) – which probed a wide range of information and insights on cities' progress in e-governance (Di Maria, Vergani and Paskaleva-Shapira, 2005). A set of 50 indicators was applied to examine issues such as general framework conditions, enablers and drivers, e-services, technology, organisations, citizens' participation and integration with national and regional processes and policies which were further linked to future technologies, government modernisation, new business models, impacts, priorities, investment strategies, decision- and policy-making processes and user demands.

b. Two consecutive studies, one that mapped the theoretical underpinnings of modelling the e-city governance readiness (Paskaleva-Shapira, 2007) and another more in-depth investigation of ongoing trends and changes in EU cities, which concluded that urban governments will need to adapt their e-governance approaches to innovation policy-making strategically so as to meet the increasing challenges of the competitive and intelligent city (Paskaleva-Shapira, 2008). In the next two sections, we take this notion forward and explore the e-governance survey results from the perspective of the four main dimensions of the smart city identified in this chapter: networks and partnerships, integrated e-services, e-participation and innovation of policy.

State of play and emerging trends

The survey research shows that information provision and e-services are the main focus of city e-governance in Europe in the present, but the overall level of progress is moderate and differs significantly between cities. The most advanced ones – such as Reykjavik, Camden, Linz, Barcelona, Siena and Venice – use a more comprehensive and consistent approach to e-governance, based on open source and standards. A wide range of sophisticated online services to citizens and businesses are provided and interactive service applications are commonly used. Residents are frequently asked to participate in local decision-making initiatives and developing online applications and services by request is a main priority of the local administrations, followed by e-consultation. State-of-the-art data and knowledge management systems are often used in the back offices and innovation appears a driving force of development. These cities have also made substantial investments in creating multiple points of access, based on ICT, most often motivated by the need to improve public awareness of the network technologies and build a closer relationship between government and citizens by reducing physical distances and increasing communication. Through multimedia or widely diffused technology solutions (mobile phones, TV), these cities generally seek to foster the use of ICT among the citizens and increase the use of online public services, often aiming for broader public goals.

Yet, learning and training programmes and multiple channels are of a lesser priority and more cohesive, integrated approaches to technology platforms are generally uncommon, while building citizen's capacity is of no significant relevance to providing integrated online services at all. The second, less advanced group of cities – such as Athens, Nice, Marseille and Manchester – have generally made moderate progress in e-governance, offering less complex e-solutions and including a modest level of participation of their users. The last and least progressed type of cities, such as Karlsruhe in Germany, are still at the onset of strategising and planning for e-governance in its complex dimensions.

E-participation, considered in regard to the citizens' access to opportunities, ICT and capacity building, as well as to the strategic priorities, appears not to be in the forefront of current e-governance emphasis. Overall, citizens are provided with a limited and unequal access to the social and economic opportunities in their communities. Only one fourth of the cities offer such opportunities online, but to just half of its population. Most of the cities are yet to make progress in this area. Some are already developing the infrastructure necessary for online offering through interactive solutions, but the majority are still to upgrade their capacities. All, however, insist on the need to move towards multiple channels for connecting with the users which will increase access and the benefits of urban e-governance.

But the impacts and the advantages of e-governance for local democracy and prosperity remain largely unfamiliar to the cities and their wider constituencies. Existing ICT systems and capacity building programmes primarily focus on web content management and less on the urban governing and management issues. Tools such as e-forums, e-discussions and e-boards for suggestions and comments from the citizens are often used by city administrations to promote involvement in local affairs. Yet, only half of the cities consider that they have really engaged with the citizens. Other e-participation tools such as online polls and e-surveys, e-voting and e-referenda are still of limited use, while more deliberative platforms such as institutions' or citizens' e-panels or local e-parliament are almost not present. Similarly, citizens' capacity building e-tools which produce content remain broadly unpopular. Future e-participation strategic priorities include, first and foremost, consultations on city development programmes, followed by tools that promote transparency (information access) and solicitation of people's opinion and suggestions on handling the urban affairs. On the other hand, online polling and voting is a long-term aim and enhancing social cohesion and people's interest in their place of living is a least considered option. Increasing public participation in decision making (influence/contribution to urban competitiveness and community well-being) is a concern of only half of the cities. Least acknowledged remain goals such as enhancing the quality and effectiveness of decision making and improving the transparency of the public sector in its workings.

With regard to the driving actors, despite ICT providers being in the forefront of delivery, local administrations recognise the key to this lies with leadership. Indeed, in the more advanced cities, the roles of the mayors and the city council have proven pivotal for innovative local e-governance, and despite acknowledging the importance of networking and collaboration, only a few cities have linked with regional and national e-government systems; one third of them have made a moderate progress; another third report an insignificant advance; while the rest, to their disadvantage, have largely failed to link with other levels of digital government.

Implications for the smart city and policy innovation

Clearly, the national and urban contexts are different for the different study cities, and the notion of e-governance has a different resonance in each one. Defining the local profiles is the top priority and it has to go hand in hand with what is essential for all actors, as the promotion of policies and practices depends on this context. Approaches to e-governance would arguably vary from city to city, but aiming for the smart city allows long-term sustainable perspectives for all. E-governance readiness is thus an outcome that is evident in the long-term development strategies of cities. There are several general trends and implications that are emerging from this study in this regard. Above all, it is evident that e-governance has to evolve as a multi-phase transformation, rather than a single-step process which involves a modernised city government serving the needs of the citizens and the well-being of the community. Networking, integration, inclusion and policy innovation emerge as the key dimensions and understanding the policy context of e-governance transpires as a strategic innovation methodology which informs the cities' approach to becoming smart. So this research demonstrates that positive and sustainable results concerning e-governance as an enabler of the smart city can be obtained through a synergic approach of policy innovation in five main directions:

Comprehensive and systematic approach to city e-governance. Overall, cities are in dire need of understanding the context of e-governance in its complexity and the impacts that it creates. Therefore, they should develop a synergetic approach to their e-governance plans and initiatives, ensuring not just the provision of services and information to their citizens, but the advancement of sustainable urban participatory processes, structures and organisation. Understanding, planning and seeking broad urban impacts is a prerequisite for becoming ready for the challenge. Recognising the benefits for social integration, inclusion and curbing the digital divide is not enough. Cities should also consider the full potentials of citizens' participation in the decision making and for urban re/development and long-term urban competitiveness. The latter, however, would have to be considered along with networking and knowledge creation opportunities to allude to the smart city.

Collective knowledge networks. Building, improving and expanding networking within the government structures and outside, with other urban organisations and players, should become a prerequisite for any implementations. Bringing in the marginalised actors, expanding beyond the ICT companies' role in designing and implementing e-governance and raising up the city authorities as the key promoters of local e-governance is critically important. Ensuring the political support and strategic policies along with the visions and leadership to motivate collective knowledge deserves a special attention.

The mayor and city government, based on the city council's directions and e-governance action plans, should strategically manage the transition. To avoid waste of resources and overlap of activities, investing priorities should trigger the emerging needs in building community knowledge networks and platforms which can integrate diverse local knowledge reflecting the wider urban contexts and replicate or develop the technical networks and infra-structures necessary. So the key organisations among the sectors and across the territory must be defined and the horizontal and vertical e-governance structures, the first linking the urban stakeholders and the second the metro-politan levels of governance such as regions and city, should be mapped out in advance. And lastly, but not least, the new business models such as public–private partnerships as the means of implementation of technology, manage-ment and organisational transformation should set up the process.

Integrate e-services. The engagement of citizens is a central issue in the development of e-governance within an urban economy. Involving the people in government services in education, health, administration, security, police and so forth is increasingly viewed as a channel for enhancing the customer perspective, which will bring about citizen-centred services which are user-relevant and focusing on the users' needs when they are required. It is appar-ent from this study that some of the benefits promised by e-governance can only bear fruit if services are made for their real users – businesses, citiess and visitors. This makes sense given the variety of actors involved, and the fact that service infrastructure is often delivered through private agencies. Therefore, a major challenge for EU city governments in the early stages of e-governance development is to identify actual users and their needs and to design services according to the identified target group together with other service providers involved. Enhanced interactivity of the services is expected to improve government delivery of services. Yet, government e-services should not be thought of as the main objective of e-governance.

E-participation. It seems that the advent of e-governance should bring about new opportunities to enhance the smart city, based on e-democracy principles, tools and applications. Little concern about e-democracy issues was found in this study, so future investments should increase the degree of interest and involvement of citizens in urban development affairs so that the benefits expected from e-governance can bear real fruit. The findings indi-cated that e-governance initiatives are still predominantly non-interactive and non-deliberative and the level of the citizens' inclusion is moderate. Providing access to the entire urban population will ensure its effective participation in developing the collective knowledge networks on the road to the smart city. But for this to happen, e-participation should be put on the city e-governance agenda and citizens-led processes should gradually evolve. Along with the existing e-consultation and online forums, other e-participation tools and applications should also be applied and community-based collaborative knowledge management platforms should be promoted. Here direct and delib-erative participation in urban decision making and planning can truly occur, so as to enable collaboration, networking and coordinated urban interaction. Building the inclusive city should thus be placed at the heart of the smart city. Inclusive decision-making processes should be promoted as a practical strategy for translating the norms of good urban governance into practice. Thus the smart city will represent both the final vision and the process used to create it. But there is obviously a need for a comprehensive framework in policy making and regulation, involving a broader mindset, particularly when it comes to use and regulation.

Policy innovation. This research has argued that e-governance is a long-term and dynamic phenomenon to smart city change. In the previous sections, we conceptualised the interface between e-governance and the smart city and explored the specific dimensions of their relationship. The analysis also brings to light the policy innovation issues that have emerged from the pressures that competitiveness, the evolution of democracy and government and the development of ICT have imposed. This co-evolutionary process goes far beyond the mere extension of services or information, as e-governance is too often understood. Rather, it is a collective process of solution-finding, involving the key stakeholders, among which the city government has to function in a multi-level and open-ended collaborative digital environment aiming for long-term urban competitiveness goals. The innovation of policies, regarded as the successful introduction of new policies that add new value and improved qualities in the areas of their application (Fagerberg, 2004), thus concerns three main areas.

First, *capacity building* should become a main driver of city e-governance by: a) involving public servants, politicians, citizens and local communities on a sustainable basis in the use of ICT through various channels – distributed points of access, easy-to use technologies and services, social programmes of inclusion and effective and efficient service responses; b) fostering collaborative knowledge management strategies within and between governments, businesses and communities to develop collective knowledge platforms of e-governance; c) promoting networking with other urban actors to deliver custom-based solutions on a competitive basis, for example in re-use of e-solutions and e-knowledge; d) investing in networking with and within cities and other local authorities to spread and re-use successful solutions and best practices and pool resources for further innovation; and e) raising awareness among politicians of the potentialities of e-governance as a key instrument to increase the quality of the relationships with citizens and to improve administration processes.

Second, *innovation progress* should be based on innovation through expansion of the innovation spectrum with the focus being placed on: a) product (e-services), a wide set of integrated online services for citizens and businesses and the entire process and the organisational reforms; b) context-based innovation with regards to new concepts, services and networks innovation; c) change innovation dynamics – more networked, more systemic, more knowledge intensive – and changing the boundaries – more local, yet global; and

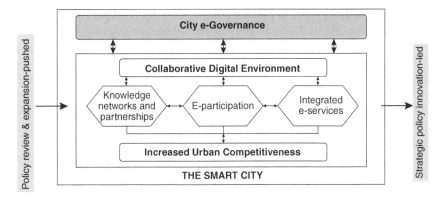

Figure 3.3 Model for e-governance enabled smart city

last but not least d) innovation as a long-term strategy and not innovative activities.

Third, *strategic policy* to review and enhance policy should become key to strategic city e-governance policy-making by: a) expanding the domain of e-governance policy; b) horizontal expansion – urban policy systems and heterogenous actors; c) vertical expansion – metropolitan, regional and international; d) inclusion of new aspects and issues – citizens' participation in urban planning and policy-making (looking outside the system – changes in the economy and society; mapping the outlook); e) city e-governance policy framework based on urban context and user demands; using governance as an innovative framework of public policy in e-governance (technology innovation, government modernisation and citizens' participation), and policies supporting implementation (leadership, innovation, strategic vision, public–private partnerships and other urban development policies enabling successful take up such as competitiveness, sustainability and participatory governing).

Concluding comments

In response to the growing pressures of globalisation and ICT, urban communities have begun to sketch out the first drafts of the 'smart cities'. But the global-local context of the smart city, i.e. its competitiveness, requires collaborative innovation and learning knowledge, amongst other approaches. Smart cities may be viewed as a template of how e-governance can allow local systems to effectively harness these global forces and deploy the results in a positive generation of new opportunities, socially, politically and economically (see Figure 3.3).

This study has demonstrated that capitalising on the benefits from e-governance requires further innovation in policy and practice in the building of the smart city. This involves the transformation of strategies into the right policy measures and practical actions (Wilks-Heeg and North, 2004). The general model proposed below can be used as a guide in such efforts.

In the quest for the smart city, networking, participation, integrated e-services and policy innovation have all emerged as key drivers of growth and essential goals of the smart city strategies. Each has a strong governance component, which requires real partnership between government, business and civil society for co-ordinating actions and adapting policies to local conditions and global demands. Cities that can shift from the traditional bureaucratic approach to using ICT (Moon and Welch, 2005) to support e-governance can achieve greater smart city objectives.

On the road to transformation, they will have to move from the predominantly non-interactive and non-deliberative stage (Chadwick and May, 2001); to the portal stage, with fully executable and integrated service delivery; and the seamless stage, with full integration of e-services across administrative boundaries; and the interactive democracy stage, with e-democracy tools which allow online engagement of the stakeholders in the process, shaping, debating and implementing service provision and public policies (Clayton and Streib, 2003; Santos and Heeks, 2003). The enhancement of urban competitiveness and citizens' well-being will be the most important benefits for the community. Dedication to that vision is reflected in strategic priorities, investment practices, capacity building and innovation pursuits.

The extent to which the use of e-governance is successful shall depend on how councils ensure that it guarantees the growth of the smart city. The right policy directions must exist, as well as the commitment to democracy. Policies should be organised to support new outcomes, processes, competences and organisational forms. A novel approach to policy as review and innovation can facilitate the shift. So a truly smart city will need to develop comprehensive strategies and approaches to address, in depth, the issues surrounding the progress and readiness of e-governance. But it will take more than yet another online service or consultation poll to change the enduring pathway to the smart city.

Acknowledgements

This research has been partially funded by the European Commission, DG/RTD Information Society Technology, Six Framework Programme, IntelCities IP Project (IST-2002–37373).

Notes

1 This is the revised version of: Paskaleva, K., Enabling the Smart City: The Progress of City e-Governance in Europe, *International Journal of Innovation and Regional Development*, 2009, 1 (4): 405–422.
2 The IntelCities Integrated Project took place during 2004–2007 and brought together 18 cities, 20 ICT companies and 33 research groups from Europe to pool advanced knowledge and experience of electronic government, urban planning and knowledge systems and citizen participation from across Europe and create a new

and innovative interoperable e-government platform and services to meet the needs of both citizens and businesses.

References

Backus, M. (2001) E-governance and Developing Countries, Research Report No.3, Retrieved from: http://www.ftpiicd.org/files/research/reports/report3.doc.

Baron, S., Field, J. and Schuller, T. (2000) *Social Capital: Critical Perspectives*, Oxford, Oxford University Press.

Barzelay, M. (2001) *The New Public Management: Improving Research and Policy Dialogue*, Berkeley, University of California Press.

Borja, J. and Castells, M. (1996) *Local and Global: Management of Cities in the Information Age*, London, Earthscan.

Bounfour, A. and Edvinsson, L. (2005) *Intellectual Capital for Communities: Nations, Regions and Cities*, Oxford, Butterworth-Heinemann.

Buck, N., Gordon, I., Harding, A. and Turok, I., eds (2004) *Changing Cities: Rethinking Urban Competitiveness, Cohesion and Governance*, University of Bristol, UK Policy Press.

Castells, M. and Hall, P. (1994) *Technologies of the World*, London, Routledge.

Centre for International Development (2000) *Readiness for the Networked World: A Guide for Developing Countries*, Harvard University, Retrieved from http://cyber.law.harvard.edu/itg/libpubs/guides/english_guide.pdf.

Chadwick, A. and May, C. (2001) Interaction between States and Citizens in the Age of the Internet: 'E-Government' in the United States, Britain and the European Union. Proceedings of the American Political Science Association Conference, San Francisco, CA, August 30–September 2, 2001.

Clayton, J and Streib, G. (2003) The New Face of Government: Citizen-Initiated Contacts in the Era of E-Government, *Journal of Public Administration Research and Theory*, 13: 83–102.

Cloete, F. (2002) Capacity Building for Sustainable Local Governance in South Africa, in Parnell, S., Pieterse, E., Swilling, M. and Wooldridge, D., eds, *Democratising Local Government: The South African Experiment*, Cape Town, UCT Press.

Coe, A., Paquet, J. and Roy, J. (2001) *E-Governance and Smart Communities: A Social Learning Challenge*, London, Sage Publications.

Cooper, I., Hamilton, A. and Bentivegna, V. (2005) Sustainable Urban Development: Networked Communities, Virtual Organizations and the Production of Knowledge, in Curwell, S., Deakin, M. and Symes, M., eds, *Sustainable Urban Development Volume 1: The Framework, Protocols and Environmental Assessment Methods*, Oxford, Routledge.

Di Maria, E., Vergani, S. and Paskaleva-Shapira, K. (2005) *E-Governance Practices, Strategies and Policies of European Cities: State-of-the art*, 'IntelCities' Project Report D11.1.2, Retrieved from http://www.intelcitiesproject.com.

Eger, J. (1997, October 26). Cyberspace and Cyberplace: Building the Smart Communities of Tomorrow [Online].

European Commission (2002) *Vision and Roadmaps for Sustainable Development in a Networked Knowledge Society*, Information Society Directorate, General Report, Retrieved from ftp://ftp.cordis.europa.eu/pub/ist/docs/ka2/visionsandroadmapsfinal20020131.pdf.

European Commission (2003) *European Governance: A White Paper*, Retrieved from http://europa.eu.int/eur-lex/en/com/cnc/2001/ com2001_0428en01.pdf.

European Commission (2005a) *e-Government Communication*, Retrieved from http:// europa.eu.int/information_society/eeurope/2005/.

European Commission (2005b) *Putting Europe High on the Global Map of Science and Technology*, communication of the Commission of the European Communities, Brussels, 24 September, Retrieved from http://ec.europa.eu/research/press/2008/pdf/ com_2008_588_en.pdf.

European Commission (2007) *Fourth Ministerial eGovernment Conference: Reaping the Benefits of e-Government*, 20–21 September 2007 in Lisbon. Retrieved from http://ec.europa.eu/idabc/en/document/7196.

Fagerberg, J. (2004) Innovation: A Guide to the Literature, in Fagerberg, J., David, C., Mowery, D.C. and Nelson, E., eds, *The Oxford Handbook of Innovations*, Oxford, Oxford University Press.

Fifth Worldwide Forum on e-Democracy (2004) Issy-les-Moulineaux, Paris, France 29–30 September, 2004, Retrieved from www.issy.com/Rub.cfm?Esp=1&Rub=19.

Frissen, P. (1997) The Virtual State: Postmodernisation, Informatisation and Public Administration, in Loader, B. ed., *The Governance of Cyberspace: Politics, Technology and Global Restructuring*, London, Routledge.

Global e-Policy eGovernment Institute (2003) *Digital Governance in Municipalities Worldwide: An Assessment of Municipal Web Sites throughout the World*. Retrieved from http://unpan1.un.org/intradoc/groups/public/documents/aspa/unpan012905. pdf.

Horan, T. and Wells, K. (2005) Digital Communities of Practice: Investigation of Actionable Knowledge for Local Information Networks, *Knowledge, Technology, and Policy*, 18 (1): 27–42.

Leadbeater, C. (1999) *Living on Thin Air: The New Economy*, London, Viking.

Loader, B, ed. (1997) *The Governance of Cyberspace: Politics, Technology and Global Restructuring*, London, Routledge.

Moon, M.J. (2002) The Evolution of e-Government Among Municipalities: Rhetoric or Reality, *Public Administration Review*, 62 (4): 424–433.

Moon, M.J. and Welch, E. (2005) Same Bed, Different Dreams? A Comparative Analysis of Citizen and Bureaucrat Perspectives on E-Government, *Review of Public Personnel Administration*, 25 (3): 243 – 264.

Noordhoek, P. and Saner, R. (2004) Beyond New Public Management: Answering the Claims of both Politics and Society, *Public Organisation Review*, 5 (1): 35–53.

Oakley, K. (2002) What is e-Governance? Integrated Project 1: *e*-Governance Workshop 10–11 June, Strasbourg.

Odendaal, N. (2003) Information and Communication Technology and Local Governance: Understanding the Difference between Cities in Developed and Emerging Economies, *Computers, Environment and Urban Systems*, 27 (6): 585–607.

OECD (2003) *Engaging Citizens Online for Better Policy-Making*, OECD Policy Brief, March, 2003, Paris, OECD Observer, Retrieved from http://www.oecd.org/ dataoecd/62/23/2501856.pdf.

Paskaleva-Shapira, K. (2007) E-City Europe: State of Play, Propositions and Opportunities, Proceedings of the International Conference *Intelligent Environments*, Ulm, Germany, September 24–25, IEEE.

Paskaleva-Shapira, K. (2008) Assessing local e-governance in Europe, *International Journal of Eletronic Governance*, 4 (4): 17–36.

Preissl, B. and Muller, J. (2006) Governance of Communication Networks: Connecting Societies and Markets with IT, Physica-Verlag.

Rakodi, C. (2001) Urban Governance and Poverty: Addressing Needs, Asserting Claims: An Editorial Introduction, *International Planning Studies*, 6 (4): 343–356.

Santos, R. and Heeks, R. (2003) *ICTs and Intra-Governmental Structures at Local Regional and Central Levels: Updating Conventional Ideas*, IDPM, University of Manchester.

Snellen, I. and Van Der Donk, W., eds (1998) *Public Administration in an Information Age: A Handbook*, Amsterdam: IOS Press.

Steins, C. (2002) E-government: The Top Ten Technologies, *Planning*, September.

Tat-Kei Ho, A. (2002) Reinventing Local Governments and the E-Government Initiative, *Public Administration Review*, 62 (4): 410–420.

Thorleifsdottir, A., Paskaleva-Shapira, K., Forseback, L., Tzovaras, D., Christodoulou, E. and Schnepf, D. (2004) Best Practices in e-Governance, *IntelCities Research Report D15.2.2*, Retrieved from www.intelcitiesproject.com.

Torres, L., Pina, V. and Royo, S. (2005) E-government and the Transformation of Public Administrations in EU Countries: Beyond NPM or Just a Second Wave of Reforms?, *Online Information Review*, 29 (5): 531–553.

Torres, L., Pina, V. and Acerete, B. (2006) E-Governance Developments in EU Cities: Reshaping Government's Relationship with Citizens, *Governance*, 19 (2): 277–302.

UN-HABITAT (2002) The Global Campaign on Urban Governance, 2nd edition, Concept paper, Retrieved from: http://www.unhabitat.org/governance.

Van Den Berg, L. and Van Winden, W. (2002) *Information and Communication Technology as Potential Catalysis for Sustainable Urban Development: Experiences in Eindhoven, Helsinki, Manchester, Marseilles and The Hage*, Aldershot, Ashgate Publishing Ltd.

Van Der Meer, L. and Van Winden, W. (2003) E-governance in Cities: A Comparison of Urban ICT Policies, *Regional Studies*, 37 (4): 407–419.

Wilhelm, A.G. (2004) *Digital Nation: Towards an Inclusive Information Society*, Cambridge, MA: MIT Press.

Wilks-Heeg, S. and North, P. (2004) Cultural Policy and Urban Regeneration, *Local Economy*, 19 (3): 305–311.

Wong, W. and Welch, E. (2004) Does e-Government Promote Accountability? An Analysis of Website Openness and Government Accountability, *Governance: An International Journal of Policy, Administration, and Institutions*, 17 (2): 275–297.

World Economic Forum (2003) *Global Competitiveness Report 2003–2004.*, USA, Oxford University Press.

Zimmermann, W. (2005) eStrategy for e-Government in Cities: An Introduction, UMP-Asia Occasional Paper No. 62, UMP Regional Office for Asia and the Pacific, Retrieved from http://www.serd.ait.ac.th/ump/OP%20Dr.%20Willie%20ed%20vers%202005%20from%20GS.pdf.

4   The IntelCities community of practice[1]

Mark Deakin, Patrizia Lombardi and Ian Cooper

Introduction

This chapter examines the IntelCities community of practice (CoP) as an organisation supporting the capacity-building, co-design, monitoring and evaluation of eGovernment (eGov) service developments. It begins by outlining the IntelCities CoP and goes on to set out the integrated model of electronically enhanced government (eGov) services developed by the CoP as a means by which to build the capacity that is needed to co-design, monitor and evaluate the IntelCities e-learning platform, knowledge management system (KMS) and digital library. The chapter goes on to examine the information technology (IT) underlying the organisation's e-learning platform, KMS and digital library, as a set of semantically interoperable eGov services supporting the crime, safety and security initiatives of socially inclusive and participatory urban regeneration programmes.

The notion of the intelligent city

The notion of the intelligent city as the provider of electronically enhanced services has become popular over the past decade or so (Graham and Marvin, 1996; Mitchell, 2000). In response to this growing interest in the notion of intelligent cities, researchers have begun to explore the possibilities of using CoPs as a means to get beyond current 'state-of-the-art' solutions and use the potential that such organisations offer to develop integrated models of e-government (eGov) services (Curwell et al., 2005; Lombardi and Curwell, 2005). What follows reports on the outcomes of one such exploration and reviews the attempt made by a consortium of leading European cities to use the intelligence that CoPs offer as the organisational means by which to get beyond current state-of-the art solutions. The CoP in question is that developed under the IntelCities Project[2] and is known as the IntelCities CoP.

The IntelCities community of practice

The IntelCities CoP is made up of research institutes, information, communication and technology (ICT) companies and cities, all collaborating with one another and reaching consensus on how to develop integrated models of eGov services. Made up of researchers, computer engineers, informational managers and service providers, this CoP has worked to develop an integrated model of eGov services and support the actions taken to host them on platforms (in this instance something known as the eCity platform) with sufficient intelligence to meet the e-learning needs, knowledge transfer requirements and capacity building commitments of socially inclusive and participatory urban regeneration programmes (Deakin and Allwinkle, 2006).

As an exercise in CoP development, the organisation is particularly successful for this reason: the intelligence it has sought to embed in cities and integrate within their platforms of eGov services is inter-organisational, networked, virtual and managed as part of a highly distributed web-based learning environment. If we quickly review the legacy of CoPs in organisational studies, the value of developing such a learning environment should become clear.

Literature on CoPs

The literature on CoPs reveals many different kinds of situated practices, all of them displaying quite varied processes of learning and knowledge gathered around distinct forms of social interaction. In this respect, Wenger's (1998, 2000) studies of CoPs concern the way that insurance claim processors and other such occupational groups learn to be effective in their job. Orr (1996) also studies the importance of CoPs amongst photocopier repair technicians. Osterlund (1996) studies CoPs as learning organisations which cut across craft, occupational and professional divisions and which transfer knowledge between them. The collective representation of CoPs in the literature suggests such organisations have certain characteristics (see Table 4.1).

Taking this representation of CoPs as a starting point for their examination, Amin and Roberts (2008) suggest there are four distinct types of inter-organisational learning and knowledge transfer: craft, professional, creative and virtual.[3] As Amin and Roberts (2008) point out, until recently it has been assumed that virtual organisations are not capable of promoting learning and transferring knowledge. However, as it becomes easier to communicate with 'distant others' in real time and in increasingly rich ways, the resulting proliferation of online learning means interest is now centering on how the knowledge dynamics of such organisations differ from CoPs dependant on social familiarity and direct engagement to sustain their mutual relationships (Ellis et al., 2004; Johnson, 2001).

Two types of online interaction

As Amin and Roberts (2008) acknowledge, there are now two types of online interaction that merit close attention as spaces where CoPs engage in learning and get involved in knowledge generation. Firstly, innovation-seeking projects

Table 4.1 Key characteristics of a community of practice

- sustained mutual relationships;
- shared ways of engaging in doing things together;
- the rapid flow of information and propagation of innovation;
- absence of introductory preambles, as if conversations and interactions were merely the continuation of an ongoing process;
- very quick setup of a problem to be discussed;
- substantial overlap in participants' descriptions of who belongs;
- knowing what others know, what they can do, and how they can contribute to an enterprise;
- mutually defining identities;
- the ability to assess the appropriateness of actions and products;
- specific tools, representations and other artefacts;
- local lore, shared stories, inside jokes, knowing laughter;
- jargon and shortcuts to communication as well as the ease of producing new ones;
- certain styles recognised as displaying membership;
- a shared discourse reflecting a certain perspective on the world.

Source: Compiled from Wenger (1998)

which can involve a large number of participants, and secondly, relatively closed interest groups which face specific problems and are consciously organised as platforms for learning about and gaining a knowledge of how to build the capacity to include 'distant others' as participants in such projects.

As they say: open source software groups provide a good example of the first CoP. Typically, they involve short-lived projects that make source code freely available to technical experts who are motivated by the challenge of solving a difficult programming problem. Successful projects of this kind are those guided by shared notions of the problem and by a core group of highly motivated experts who associate with one another to learn about the subject and transfer the knowledge generated to distant others.

More recently, however, we have seen the development of the second type of CoP. These are established explicitly by professional, expert and lay people to advance knowledge. Typically, they involve experts interested in developing and exchanging best practice, or lay people wishing to not only learn or transfer knowledge about a given subject, but build the capacity for such developments to take place via electronically mediated communication. Here a CoP is seen to emerge once the technologies for the virtual organisation are available to learn about the subject in question and success is registered in terms of the ability such platforms have to transfer knowledge.

It is also stressed that with this type of CoP, the technology available to support the development of such virtual learning organisations is something which has to be managed. That is, and as Josefsson (2005) points out, managed in accordance with a 'netiquette', where semantically rich language is used to develop a culture of engagement, replete with humour, empathy, kindness, tact and support. This way virtual learning organisations are seen to replicate the rich texture of social interaction normally associated with CoPs marked by high levels of inter-personal trust, reciprocity and, therefore, collaboration built around strong professional and occupational ties.

Defining features of the IntelCities CoP

Made up of both open source software groups, experts and lay people, the IntelCities CoP is unique in the sense that its network provides an example of a virtual organisation set up to manage the learning needs and knowledge requirements of a technological platform. Set up under the name of the IntelCities Project, the CoP offers the means:

- to meet the learning needs, knowledge transfer requirements and capacity building commitments of the organisation;
- to co-design them as a set of services that are socially inclusive and participatory and that allow users to learn about the availability of such services, how to access them and the opportunities they offer everyone to become engaged with and get involved in meeting the knowledge transfer requirements and capacity building commitments of their urban regeneration programmes;
- to monitor and evaluate the outcomes of such actions.

There are three features that define the IntelCities CoP and that give it meaning and a sense of purpose. These can be paraphrased as: building the capacity for a shared enterprise, the co-design of the online services that this joint venture creates and both the monitoring and evaluation of the developments.

Building the capacity for a shared enterprise

It is the CoP's e-learning platform that makes it possible for the online services under development to be integrated with the knowledge transfer and capacity-building technologies that are needed for this to work as a shared enterprise. This is because this platform alone makes it possible for the citizens, communities and organisations in question to collaborate with one another and build consensus on the competencies, skills and training needed for the development of the required online services.

Such a shared enterprise is made possible because:

- the ICT-enabled networks underpinning all of this are innovative in developing an e-learning platform based on open-source technologies interoperable across online services;

- in satisfying the need for a formal learning community, this high-tech, digitally enabled network allows for the planning, development and design of the online service requirements;
- such services, in turn, allow the applications under consideration to be integrated with the e-learning, knowledge transfer and capacity-building technologies supporting the regeneration programmes under review;
- such technologies allow citizens and communities to collaborate and build consensus on the competencies, skills and training needed for the online services under development to support the quintessentially civic values of this regeneration programme;
- together these networks, innovations and partnerships create the trust needed to engage citizens and show how the active participation of communities in digitally inclusive regeneration is not only intelligent, but smart in developing the social capital – norms, rules and civic values – of the ecological integrity and equity underlying this modernisation;
- here the ecological integrity and equity of the democratic renewal take the form of consultations and deliberations in government- and citizen-led decision making that engages citizens as members of a community which participates in the governance of this ongoing modernisation.

The resulting platform supports the distribution, storage, retrieval of learning material, skill packages and training materials needed for virtual organisations of this kind to bridge the digital divides that currently exist, build the capacity that there is for inclusive decision making and transfer the knowledge required for citizens to bond with one another as members of a community.

The co-design of online services

The IntelCities CoP has sought to co-design such eGov services by overcoming the limitations of a customer-focused approach and by supplementing this with a more user-centric strategy (Lombardi et al., 2009). Here collaboration is based not only on notions of either a sovereign consumer or customer, but on the consensus built between those citizens who participate in such a process of co-design (Berger et al., 2005). Here the informational and transactional logic of mass customisation is seen as being supplemented by a process of participatory co-design that is more democratic in the way it goes about meeting personal preferences. This strategy advocates that citizens participate in the co-design of products, not as customers, but as users of the services and through their involvement in workshops which promote the authoring, self-documentation and recording of their creative experiences. Here the objective of the co-design strategy is not the mass customisation of products, nor personalisation of service provision, but collectivisation of the process. In particular, the collectivisation of the process in ways that allow citizens to collaborate with one another as a community of subjects who are sufficiently empowered to govern such developments (Nikolaus et al., 2008, Binder et al., 2008).

Monitoring and evaluation

For those involved in the co-designing of eGov services, it is not so much their customisation, or the multiple channels of communication this opens up, but the outcomes of the development process that are subject to monitoring and evaluation. This is because for such a group of stakeholders, co-design provides a basic measure of value and demonstrates not whether a service can be transacted, but if it is useful. Not in terms of whether service developments can be linked to the transactional-based logic of customer exchange, but if in doing so they can also be connected to the user-centric reasoning of social need.

Bridging the gap that exists between the Type 1 and 2 organisations

As the defining features of the IntelCities CoP, terms such as capacity-building, shared enterprise, co-design, online services, monitoring and evaluation align closely with the characteristics previously highlighted by Amin and Roberts (2008) and already set out in Table 4.1.

Table 4.2 underlines the importance of these as characteristics and adds another eight that have been exploited by the network to develop a virtual learning organisation capable of bridging the gap that exists between the Type 1 and 2 (innovation-seeking and knowledge-generating) classifications offered by Amin and Roberts (2008). In what follows we should like to suggest the extra characteristics are those needed to span the divide between the transactional-based logic underlying the customisation of Type 1 and the user-centric reasoning that Type 2 adopts to support the co-design of eGov services.

In line with current definitions of CoPs as shared enterprises, the additional features clearly highlight these particular qualities and reflect their importance, but in addition to this, they also serve to underline the combined technical and social purpose of the virtual organisation in question. This suggests that in building the capacity to co-design eGov services, it is not possible for intelligent cities to develop as either Type 1 or 2 CoPs and this is because they have to be technical and social in equal measure. In other words, rest on the transactional-based logic of (innovative seeking) customisation and the user-centric reasoning of (knowledge-generating) eGov services this also gives rise to in equal measure.

The following examination of the IntelCities CoP shall to a large extent reflect this position. It shall begin by examining the capacity that the CoP has built to co-design an integrated model of eGov services and IT underlying the eCity platform developed as an intelligent solution for the virtual organisation's learning needs and knowledge transfer requirements. The examination shall then reflect on the search for an intelligent city solution

Table 4.2 Defining characteristics of the IntelCities CoP

- sustained mutual relationships;
- shared ways of engaging and in doing things together;
- the rapid flow of information and propagation of innovation;
- absence of introductory preambles, as if conversations and interactions were merely the continuation of an ongoing process;
- very quick setup of a problem to be discussed;
- substantial overlap in participants' descriptions of who belongs;
- knowing what others know, what they can do and how they can contribute to an enterprise;
- a shared discourse reflecting a certain perspective on the world;
- *shared enterprise between research institutes, ICT companies and cities;*
- *joint venture commitment to product development;*
- *building the capacity for ICTs to be used as a means of bridging the digital divide;*
- *the co-design of services;*
- *shared commitment to social-inclusion and participatory urban regeneration programmes as a means to close the gap between the information-rich and poor;*
- *support for the modernisation of local government service provision using technological platforms;*
- *consensus-based decision making, consultative and deliberative in nature;*
- *monitoring and evaluation.*

---

in terms of the step-wise logic adopted to meet the challenge that the learning needs and knowledge transfer requirements of such virtual organisations pose. From here the e-learning platform, knowledge management system and digital library developed for such purposes shall be outlined. Having done this, attention shall turn to the innovative features of this platform, management system and library and the semantically interoperable qualities of the learning, knowledge and repository services that this offers shall be reviewed. From here the examination reviews how the learning, knowledge management and digital library services now available as eGov services are integrated into the eCity platform and made available over the web.

This turns attention to what is termed the eTopia demonstrator developed to illustrate the functionality of the semantically rich eGov services in question. This term is borrowed from Mitchell's (2000) account of intelligent cities as e-topias and as organisations that are 'SMART', lean, mean, green software systems, driven by networked communities which are virtual (see Deakin and Allwinkle, 2007a; Deakin, 2007). Organisational characteristics which, the authors would add, are built on the learning needs, knowledge management requirements and digital libraries of electronically enhanced government services and that are available on eCity platforms as pools of integrated eGov services.

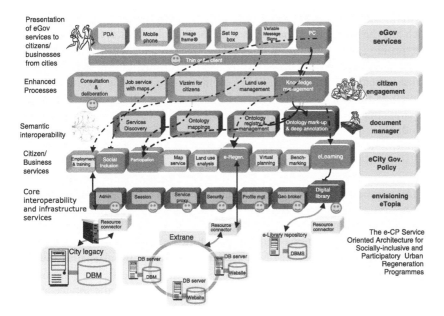

Figure 4.1 Integrated eGov services model

The integrated model of eGov services

Figure 4.1 outlines the integrated eGov services model developed by the CoP. At the front end there are a range of eGov services under development, highlighted as social inclusion, participation and regeneration, and shown in terms of the middleware integrating them between the front-end presentation tier and back-office core interoperability and infrastructure service layer of the eCity platform. This also illustrates the services located in the back office and the relationship this develops between the organisation's e-learning platform, KMS and digital library.

This shows that it is the middleware of the eCity platform that integrates the front-end delivery of government services to citizens with the back-office business functions. Figure 4.1 also shows that it is the middleware that in turn provides the opportunity for the e-learning platform, KMS and digital library making up the back-office functions to do the same and become an integral part of the eCity platform, supporting the pool of eGov services that are available for citizens to access at the front end. This integration of the e-learning platform, KMS and digital library into the middleware and use of it as the eCity platform supporting the presentation of eGov services to citizens at the front end is the challenge the IntelCities CoP has set out to meet and the customisation it has sought to co-design a solution for.

The IT underlying the intelligent city solution

The main challenge for the IntelCities CoP has been that of finding a solution which has the intelligence that cities need for the information technology (IT) underpinning the presentation of eGov services to be extensible, flexible and also have the capacity to carry existing local government legacy systems. The services oriented architecture (SOA) of the enterprise-wide business model adopted as the joint venture vehicle for such an 'intelligent solution' meets this challenge by offering the IntelCities CoP a distributed, web-based and extendable access system. This intelligence in turn offers cities the opportunity to build a web services enabled platform of eGov services, with XML IT utilisation and SOAP communication.

An important element in the initial system design relates to the use of the unique modelling language (UML) and rational unified process (RUP) methodology used for developing the integrated model of electronically enhanced government (eGov) services. This allows for the development of complex 'N-tiered' systems and the possibility of cities hosting eGov services on e-learning platforms, KM systems and digital libraries utilising the intelligence such IT offers. This has the advantage of offering a homogenous platform solution supporting the development of specific service applications meeting the e-learning needs, knowledge transfer requirements and capacity building commitments of the IntelCities CoP. It also manages to do this while leaving open the possibility for this customisation of services to be co-designed by other organisations not yet integrated into the eGov services model and eCity platform supporting this particular organisation's e-learning needs, knowledge transfer requirements and capacity building commitments.

The search for an intelligent solution

The search for an intelligent solution to the e-learning needs and KM requirements of the eCity platform has progressed by applying a stepwise logic to the challenge it poses the IntelCities CoP. This has taken the following form:

- a survey of user learning needs;
- an analysis of the knowledge requirements;
- a review of the learning and knowledge services that leading city portals provide;
- the benchmarking of existing e-learning platforms against the user's knowledge transfer and capacity building requirements;
- selecting the e-learning platform able to meet these requirements and to develop as a KM system supported by a digital library;
- integrating the aforesaid into the IntelCities middleware as a platform of eGov services delivered to citizens at the front end.

Following this stepwise logic has meant focusing attention on the underlying pedagogical issues, the competencies, skills and training requirements of IntelCities. The next step involved a review of the learning services that leading city portals offer as legacy systems and the benchmarking of the e-learning platforms these systems are based upon against the knowledge transfer and capacity building requirements of the IntelCities CoP. Here the learning services of five leading city portals were reviewed (Deakin et al., 2004; Campbell and Deakin, 2005). These included the learning services provided on the city portals of Edinburgh, Dublin, Glasgow (Drumchapel), Helsinki (Arabianranta and Munala) and Reykjavic (Garoabaer). The review found:

- the said city portals provide learning services for citizens;
- these portals provide citizens with a community grid for learning;
- much of the data available to the community is informative, telling citizens about learning opportunities in their neighbourhoods and providing links to the service providers;
- while being used by up to 10 per cent of the population and offering free email and storage, most of the services provided by the city portals are insufficiently engaging for citizens to use them as grids for communities to base their development of learning partnerships on.

As legacy systems, the review found these e-learning platforms were primarily informational, and while offering interactive learning opportunities, were insufficiently developed to meet the knowledge transfer requirements of the IntelCities CoP. However, on a more positive note, the review made clear the focus of the IntelCities e-learning platform should be the needs of the citizen; their knowledge requirements and the technology adopted to deliver this ought to break with the tradition of existing city portals, be more socially inclusive and offer greater opportunity for communities to participate in their development. With this in mind, the examination went on to benchmark the e-learning systems on which existing portals are based and examine them against the knowledge transfer and capacity building requirements they set (see Table 4.3).

The e-learning platform

Table 4.3 illustrates the results of this benchmarking exercise, presenting the average percentage scores of tools provided by 67 commercial e-learning platforms, and comparing these against the industry standard (Web CT) and European Dynamics' OSS eOWL system. This benchmarking exercise has in turn produced an OSS (open source standards) approach to e-learning, where the exercise is driven by a small e and a capital 'L'. This has opened up the opportunity to get beyond the tendency for city learning portals to merely provide links to resources held elsewhere and provided the means to customise

Table 4.3 Results of the e-learning platform benchmarking exercise

Learner tools	Commercial platforms[1]	WebCT[2]	IntelCities platform[2]
Communication Tools	57%	71%	86%
Learning Tools	62%	60%	60%
Learner Involvement Tools	64%	75%	100%
Administration Tools	79%	75%	100%
Course Delivery Tools	72%	80%	100%
Course Design	56%	83%	83%
Hardware/Software	70%	80%	63%
Pricing/Licensing	80%	40%	100%

Source: Deakin et al. (2004)
1 Indicates average percentage of learner tools covered by the 67 commercial e-learning platforms surveyed. This survey includes those used by Edinburgh, Dublin, Glasgow (Drumchapel), Helsinki (Arabianranta and Munala) and Reykjavic (Garoabaer) and approximately 60 others.
2 Highlights the percentage of functionality of individual learning tools covered by services available on WebCT and European Dynamics' OSS (e-OWL) platform

an e-learning platform capable of meeting the particular knowledge transfer requirements of the IntelCities CoP.

The learning management system

The learning management system (LMS) developed for such purposes lies at the centre of the platform. This management system provides the common ground between course tutors, trainers and learners, a virtual space where they can co-operate with one another by sharing experiences and offering personal and confidential advice on the available courses, content and communication tools. It is designed as a set of modules in which tutors can create content, administer the resulting course and create assessments for learners, while learners are able to work with that related material. The services offered by the LMS are underpinned by a set of repositories holding information on the personal data of registered members, learners' profiles, material available to support the structured course of studies and other unstructured data also available to learners.

Figure 4.2 offers a sample extract of the learning material held on the LMS. This material is drawn from the repository, matched to the personal data of registered members and distributed in accordance with the given learner profile. In this case the learning material is drawn from the LMS to support the Level 1 (Lesson 3) course of studies on eCitizenship.

The system architecture rests on three levels, each supported by a dedicated administrator: the platform administrator, the administrator and the course coordinators, tutors and trainers. Here the administrator is responsible for managing the directory of members registered to a course (this provides

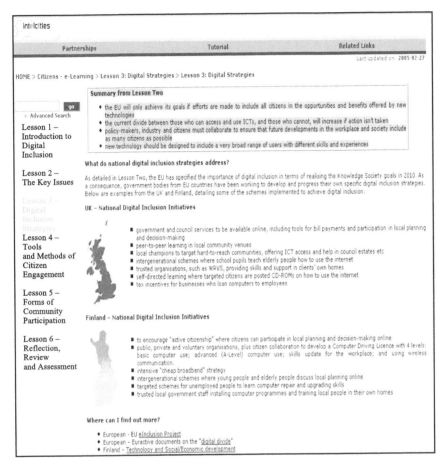

Figure 4.2 Sample of learning material for the Level 1 (Lesson 3) eCitizenship course

the interface between the course provider and the learner), whilst the tutor/ trainer will be the course content creator, and the coordinator is responsible for distributing the course(s) to the learner and the services supporting the relating studies. This is supported by core services which provide the learning content, communication, collaboration, assessment and administration of the IntelCities courses (i.e. the learning materials, skill packages and training exercises used for developing socially inclusive and participatory urban regeneration programmes) which are available to the CoP (see Figure 4.2).

The e-learning materials and courses

The e-learning materials are made up of three IntelCities courses. The first short course is aimed at members of the public with an interest in becoming more involved in civic life via the use of new technologies. The second course targets administrators within the public sector: those responsible for meeting citizens' expectations, in terms of access to electronically enhanced eGov services. The third is aimed at policy-makers and strategists within city administrations who want to make their cities leading examples of the digitally inclusive knowledge society. Together, these three courses make up the CoPs eCitizenship module. Under this heading, the course materials tackle the same core concepts: digital inclusion, citizens' expectations and the means by which cities can meet the needs of their e-ready citizens, whilst enabling access for those currently excluded from the knowledge management systems and digital technologies underlying the public's use of online services. The pitch and tone vary accordingly across the suite of materials, yet each progresses the learner towards an understanding of the tools and methods currently available to engage citizens as members of an online community.

Whilst the short course on digital inclusion provides a set of 'taster' sessions on citizens' engagement with such technologies, no prior experience of ICTs is needed as a prerequisite for the learning. It is designed to be open to everyone and provide universal access as a bottom line for the learning experiences to follow. Level 2 is targeted at citizens with different levels of experiential learning and, therefore, abilities. Those collaborating on the development of learning materials for Level 2 have developed three representative e-service users, each with different levels of familiarity with ICTs. The novice user is characterised as a citizen with little experience in using computers or the internet, but an interest in learning how to find information and pay bills online. The semi-skilled, or intermediate level user is a citizen with regular access to a computer and average-to-good ICT skills. At this level of ICT ability, the citizen is interested in locating detailed, up to-date information online and in submitting comments and feedback to the city. The advanced user has frequent access to ICTs and is highly skilled and confident in their ability to interact using the internet. This user wants maximum benefit from new technologies and is keen to interact with the city via services such as online debates and e-petitions.

These three characterisations serve to elicit the relationship between citizens' ICT skills and competencies and the e-services they expect their cities to provide. Figure 4.3 summarises this relationship. The left-hand column details the expectations of novice ICT users, the challenges these represent and the action cities can take in response to them. With little access to ICTs, such as home PCs or 3G mobile phones, the novice ICT user has little confidence in the e-services under development and the potential benefits they offer. In terms of their priorities, citizens at this level prioritise the accessibility of new online services: can they locate them easily. The challenge cities are faced with is that of meeting these very basic requirements without alienating those which have higher skill levels.

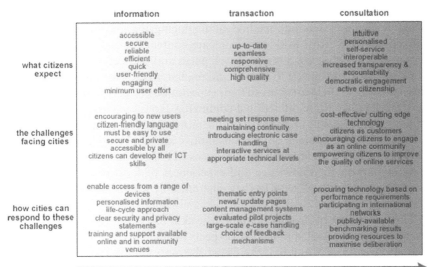

Figure 4.3 Citizens' skills and competencies
Source: Campbell and Deakin (2005)

As Figure 4.3 indicates, citizens with minimal ICT skills are unable to make use of cutting-edge, interactive technologies. Digitally excluded citizens, often amongst the most socially deprived, risk being further excluded for the reason they lack the skills to progress in the workplace and are not members of the online communities, where citizens and their cities consult with one another and meet to deliberate on issues of public concern. By investing in community-based training initiatives and online user support, cities ensure that citizens with little or no ICT experience are offered the chance to develop their skills, be included and participate in the customisation and co-design of more complex interactive online activities.

Figure 4.3 identifies citizens at the lowest skill level as seeking engagement at an informational level. Citizens who have progressed beyond basic ICT skills are referred to as seeking engagement at a transactional level. The semi-skilled, or intermediate level, user has better access to ICTs than the novice and is already comfortable accessing basic information and making bill payments online. It also identifies the intermediate user's expectations: for up-to-date information of a high quality and the seamless transition between different online services and websites.

As with the novice user, the user at this level requires services that are pitched at the appropriate skill level, again presenting the city with the challenge of meeting the needs of a diverse society. At this 'transactional' level, citizens are interested in establishing online communication with the city, and in order to engage these users and encourage repeated use of such services, cities are required to respond within set times. Electronic case handling is listed in Figure 4.3 as one method of managing the information flow and building citizens' trust in e-services, as are content management systems to ensure continuity across a range of web pages and services.

Citizens with advanced ICT skills and regular access pose an additional set of challenges to their cities, given their expectations of personalised and intuitive services like those offered in e-commerce. However, citizens at this level of ability are also able to make use of the more complex technologies that cities can offer to encourage online consultative and deliberative participation. By engaging increasing numbers of citizens in online dialogue, city administrations harness the knowledge and experiences of local people in order to improve the quality of services they customise by way of and through their co-design.

The Level 3 set of lessons examines the skill bases and competencies of a user who has just such abilities, who expects their city to provide personalised and intuitive services and to make use of the more complex technologies that cities can offer to encourage online consultation and deliberative participation. Level 3 provides a set of lessons on how cities can use the skills and competencies their citizens have to make use of these complex technologies and become leading examples of the IntelCities CoP. Two interactive video lessons have also been produced to support this set of lessons.

The pedagogy

The pedagogy of the course materials is grounded in the transformational logic of situational learning, very much action orientated and problem based in the sense that the platform's knowledge transfer capacity is framed in 'structured query language' (SQL) protocols. This can be classified as follows:

- for the basic level of learning, it is instructional, providing an outline of the material they need to be informed about and developing the literacy required for any such communication;
- with the intermediate level of user, the pedagogy is again instructional, but the emphasis here is on the social context of the eCity platform and sets out the skill bases, competencies and training needed for citizens to use the services and engage with others by carrying out online transactions, or by consulting with others as members of a community;
- the pedagogy of the advanced learner is constructivist. Drawing upon the learning of the previous level, this course uses this knowledge as a platform for citizens to use as a means of intervening in decision-making processes, engaging in consultations and deliberating with others to influence the level of government service provision. Here, users of the eCity platform learn how to actively participate as members of an online community that seeks to democratise decision making and develop the degree

of reciprocity that is needed to build trust between citizens and the organisations governing the delivery of services.

Having established the user requirements and found an e-learning platform
to carry them, attention has turned to the development of this into a KM
system and digital library supporting the activities of the IntelCities CoP.
Developed as back-office functions, attention has subsequently been given to
integrating the KM system and digital library into the IntelCities middleware
and delivering the resulting pool of eGov services to citizens wanting to learn
about them.

The knowledge management system and digital library

The KMS is organised and grouped according to the requirements of a pre-
specified, but evolving, eGov services ontology. The overriding objective of
the IntelCities CoP is to provide an e-learning platform that allows access to a
KMS and which is both accessible and usable. This objective has been met by
developing the KM system's document manager (DM). The DM has built the
capacity to perform ontology-based annotation in semantic web for the easy
creation, application and use of semantic data. This is particularly important
where learners require the KMS to perform a deep and semantically rich
annotation of materials.

The digital library is the electronic repository storing the information
available for extraction by the KMS. The rationale for developing the digital
library as part of the KM system lies with the potential the DM has to function as a service capable of:

*   capturing, storing, indexing and (re)distributing the learning materials,
    skill packages and training manuals;
*   extending this to include the formal semantics (metadata, knowledge) for
    the retrieval and extraction of the said materials, packages and manuals
    available to support the integrated modelling of eGov services;
*   offering access to the extensive range of products stored as knowledge
    objects in the digital library and available for extraction by those man-
    aging the development of the middleware as a platform for pooling the
    said eGov services together and extending delivery of them to citizens as
    front-end users.

Semantically interoperable eGov services

Utilising the semantic web paradigm, the e-learning platform is capable of
delivering data to its users in a way that enables a more effective 'query-
minded' discovery, integration and reuse of the knowledge that can be accessed
from the digital library. Through the platform's utilisation of semantic web
technologies, data uploaded by the KMS (as information available from

the system's DM) presents knowledge products codified in ways that not only correspond to documents (web pages, images, audio clips, etc., as the internet currently does), but more pre-defined objects, such as people, places, organisations and events which are deposited in the digital library. Using a pre-defined ontology of this type, the DM allows multiple relations between objects to be created.

It is here that the IntelCities project achieves its goal of integrating Type 1 and 2 CoPs and builds the capacity for the organisation to be both innovative seeking and knowledge generating. Both innovative seeking and knowledge generating in the sense in which the organisation is able to co-design eGov services as semantically interoperable codifications of the objects they relate to. Currently none of the e-learning platforms forming the basis of the CoP's SWOT analysis offer such services. Until now it has only been common to see references to the possible convergence of e-learning platforms, KMS and digital libraries. This platform and system gets beyond calls for the convergence of such technologies and begins to integrate eGov services with the ICTs available to achieve this. Perhaps most importantly of all, the outcome of all this is an e-learning platform, KMS and digital library with the embedded intelligence that cities need to deliver semantically interoperable eGov services and meet this requirement as a standard measure of the socially inclusive and participatory urban regeneration programmes that the IntelCities CoP has a particular interest in.

Integration into the eCity platform

Figure 4.4 illustrates how the IntelCities CoP proposes the integration of the e-learning platform, KMS and digital library should take place and shows the workflow supporting this. This shows the workflow as having its basis in the digital library and KMS of the e-learning platform. It also shows the workflow between the courses held on the said platform and KMS. Here the system's DM is shown to semantically annotate the learning materials, skill packages and training manuals supporting the courses held on the platform and mark them up in line with the index and classification of the eGov services ontology evolving to manage the knowledge drawn from the digital library.

These back-office functions in turn lead to the creation of the citizen engagement matrix, designed as a semantically rich grid, allowing communities to be inclusive and actively participate (via consultative and deliberative operations) in the development of the middleware as applications which this platform of eGov services delivers to the front end. These developments provide the knowledge management toolkit. This term is preferred to 'e-learning platform' because this best captures the contribution that the tools – the electronic repository, document manager, semantic annotation and mark-up system – make to the type of knowledge management and digital library services currently found on city portals.

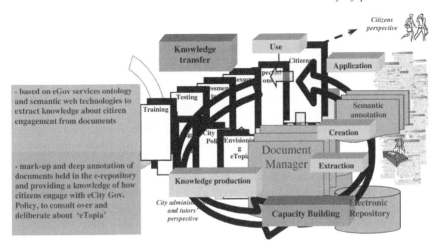

The following labels appear within the figure:

- based on eGov services ontology and semantic web technologies to extract knowledge about citizen engagement from documents

- mark-up and deep annotation of documents held in the e-repository and providing a knowledge of how citizens engage with eCity Gov. Policy, to consult over and deliberate about 'eTopia'

Knowledge transfer · Use · Citizens perspective · Citizen · Application · Testing · Training · Semantic annotation · Creation · City Policy · Envisioning eTopia · Document Manager · Extraction · Knowledge production · Knowledge production · City administration and tutors' perspective · Capacity Building · Electronic Repository

Figure 4.4 Integration into the eCity platform

While it is recognised this journey from the front-end eGov services to the middleware and towards the back-office functions represents a significant detour, it is undertaken because the path taken does mark a significant step forward. Not only in terms of the additional learning services that existing city portals are now able to offer, but in 'squaring of the circle' and providing a platform with the intelligence – KMS and digital library – to integrate the front-, middle- and back-office sections of their organisation as a virtual CoP. That is, as a CoP which in this instance is based on standards that are interoperable across a growing pool of extensible eGov services and which have the capacity, customisation qualities and co-design features to support socially inclusive and participatory urban regeneration programmes.

The 'eTopia' demonstrator

At present this integration is mainly technical, concerning the software developments needed to host such services and meet with the semantics of the platform's e-learning needs, knowledge transfer requirements and capacity building commitments. This currently takes the form of an 'eTopia' demonstrator, showing in 'session-managed logic' how the eCity platform accesses the extensive pool of eGov services located in the back office and uses the intelligence embedded in the middleware to deliver Level 3 (advanced e-Citizenship) courses on the consultative needs and deliberative requirements of such developments. This provides a 'real time' demonstration of the platform's capacity to be 'SMART' in developing both the technical and semantically rich content required for the middleware to begin supporting the socially inclusive consultations and participatory deliberations of urban regeneration programmes. These enhanced processes of consultation and deliberation also have the advantage of offering citizens multi-channel access to such eGov services, customised, co-designed and bundled together as socially inclusive and participatory urban regeneration programmes for bringing about improvements in the quality of life (Deakin and Allwinkle, 2007b). This goes a long way to:

- uncover the business logic needed to base the intelligence-driven (re)orga-nisation of cities on, and standards required to benchmark the perfor-mance of the platform against;
- provide the performance-based measures needed to assess whether any plans that cities posses to develop eGov services (over the platform) have the embedded intelligence (the learning, knowledge-based competencies and skills) required to support such actions;
- also provide the means to evaluate if such planned developments build the (intellectual) capacity – learning, knowledge-based competencies and skills – needed to support such actions.

Testing the demonstrator

In addition to developing the semantically interoperable eGov services, the IntelCities CoP has also sought to evaluate how well they perform as components of the eCity platform. This has meant developing three 'eTopia demonstrator' storylines, where the typical learners referred to previously use the eCity platform to query the development of urban regeneration programmes by either searching for information on a given initiative, gaining access to possible online transactions supporting any such actions, or by getting involved in the customisation and co-design of those consultations and deliberations underlying the governance of such proposals.

The three storylines developed scenarios for:

- accessing local services in neighbourhoods subject to regeneration;
- carrying out online transactions related to the use of land;
- consultations and deliberations about the safety and security issues underlying the governance of urban regeneration programmes.

These storylines aim to fulfil three requirements: first to continue the loosely structured scenarios used to demonstrate the significance of the citizen-led learning agenda developed under the alpha version of the eCity platform; secondly, to integrate this into the back-office business logic of the beta version testing of the eCity platform; and thirdly, to establish whether the interoperability resulting from the vertical and horizontal integration of the services is beneficial because it enables urban regeneration programmes to work better in meeting citizens' expectations. The following shall summarise the scenario-based testing of the third and 'advanced' level of eCitizenship held on the e-learning platform and accessed via the KMS.

The exercise began by introducing an 'integrated eGov services scenario' in which two people, Mark and Sarah, are keen to discover what governance services the eCity platform offers for them to learn about how it is possible to become actively involved in initiatives promoted by cities to tackle problems associated with crime in their neighbourhood. The material demonstrates the ways in which Mark and Sarah can use the eCity platform (i.e., e-learning platform and KMS) to not only learn about what they can do to tackle crime, but gain a knowledge of how the community's participation in such initiatives can lead to the development of safe and secure neighbourhoods.

The integrated eGov services scenario

Both Mark and Sarah feel their family and work commitments have prevented them from becoming more involved with local groups in the past. However, both are keen on home computing and have broadband connections to the internet. Mark feels that the city's website should provide information on crime rates and proposes that he and Sarah should log on and initiate a search to see how much they can learn about crime prevention initiatives online. They both want to know what their local administration is currently doing to address neighbourhood issues across the city and to submit their comments on past and present initiatives. They also feel it would be valuable to see what local groups are doing to tackle crime and whether any operate in their neighbourhood. They are also keen to discover how they, as citizens, can use the platform of services available on the city's information portal to ensure the urban regeneration programmes affecting their neighbourhoods are effective in tackling crime and making the areas safe and secure.

The steps

The steps Mark and Sarah can take to use the eCity platform as a means to begin tackling the problems they encounter are set out in Table 4.4.

Mark's work frequently takes him to one of the country's larger cities and he has been impressed by local initiatives to address neighbourhood issues, such as a 'Neighbourhood scheme involving local residents'. He is also interested in comparing the crime rates in his neighbourhood with those in other cities and finding which crime prevention schemes seem to work best.

The information flow

The flowchart (see Figure 4.5) demonstrates how the eCity platform helps Mark and Sarah to query the developments they have a particular interest in and use this to find the information they need. They are able to access a wide range of data sets from their local administration, such as policy documents

Table 4.4 Step-wise logic of the service discovery

Step 1: Use the city's website to view information on current neighbourhood policies, strategies and targets;

Step 2: Use the website to access a list of current public consultations;

Step 3: Use the search tools to learn about any local online crime prevention and environmental clean-up groups, run either by local people or by the city;

Step 4: Use the neighbourhood reporting service on the interactive maps to problems such as abandoned cars, graffiti and fly tipping;

Step 5: Set up a web page for local people interested in tackling crime, security and environmental problems;

Step 6: Post comments on the city's discussion boards;

Step 7: See how the city compares to other cities on issues like crime and pollution;

Step 8: Submit a formal e-petition, setting out an agenda for tackling these types of neighbourhood issues.

and strategies but, most importantly, they are able exploit the potential to use this information to interact with other like-minded people as part of a larger group.

For this purpose, Mark and Sarah can develop a web page and host it on the city's learning platform, setting out their concerns about crime and encouraging others to join them as members of an online community discussing how the city should tackle neighbourhood safety and security issues. As an online community they are also able to compare their city's agenda for tackling crime with those of other administrations and learn about good practice examples from elsewhere. These materials can in turn be used to shape the community's online discussions and enable Mark and Sarah to submit a formal e-petition to those responsible for leading the development of such initiatives.

Meeting citizens' expectations

The results of this testing exercise are encouraging. Figure 4.6 demonstrates the responses of the group in question. As this shows, all found the scenarios, steps and information flow to be understandable, in terms of the vocabulary used, and also easy to follow. One attendee commented, 'I found the material quite open, easy to understand. It made me think more about how I would go about things in the future.' As Figure 4.6 also illustrates, most of those participating in the testing exercise found the demonstration to offer a useful representation of how to learn about the eCity platform's online services and use the information this uploads to transfer knowledge about how communities of like-minded people can ensure that the safety and security measures of urban regeneration programmes work in their interests.

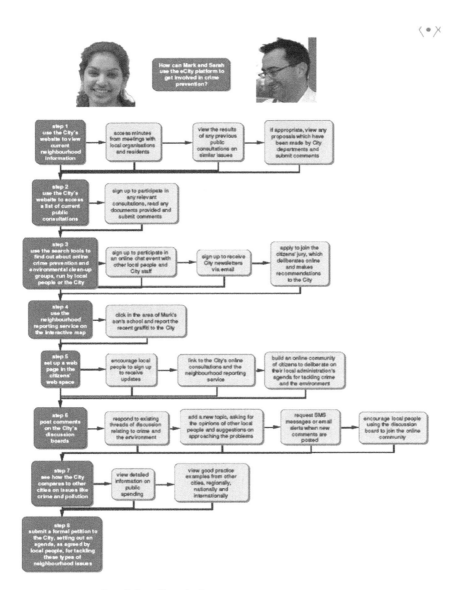

Figure 4.5 Benefits of the eCity platform

Conclusion

Made up of both open source software groups, experts and lay people, this chapter has argued that the IntelCities CoP is innovation seeking in the sense that the network provides an example of a virtual organisation whose customisation is co-designed to manage the learning needs and knowledge-generating requirements of a technological platform.

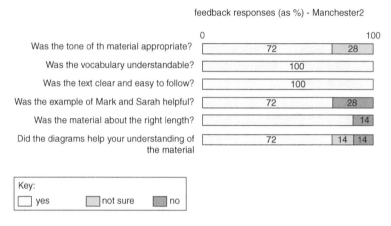

Figure 4.6 Feedback responses to the scenario

The examination has suggested there are three features that define the IntelCities CoP and which give it a sense of meaning and purpose. These are: building the capacity for shared enterprise, the co-design of online services and both their monitoring and evaluation. It has gone on to underline the importance of these as characteristics of the IntelCities CoP and in this aim has added another eight qualities that have been exploited by the network to develop a virtual learning organisation that is capable of bridging the gap that exists between the type 1 and 2 classifications of knowledge generation offered by Amin and Roberts (2008). This has been done in the interests of illustrating how the IntelCity CoP spans the divide between what are in crude terms representative of the transactional-based logic and user-centric reasoning of eGov service provision. The additional features referred to clearly highlight these qualities and reflect their importance, but in addition to this they also serve to underscore the technical and social purpose of the virtual organisation responsible for the customisation and co-design of the electronically enhanced service provision in question.

This suggests that in developing integrated eGov service models it is not possible for intelligent cities to develop as either type 1 or 2 CoPs because as virtual learning organisations, the shared enterprise and joint ventures they enter into as a process of customisation means that online service developments have to be co-designed in ways that are both innovation-seeking and which are also knowledge-generating. Having gone on to discuss the technical solutions adopted to integrate the eGov services model with the legacy systems operated by cities involved in this enterprise, attention has turned to the SOA adopted by the venture as the business model for the eCity platform.

As has been shown, these developments are valuable because they provide the means to address the criticisms of the learning services currently available on city portals and offer the opportunity for the emerging technologies of the e-learning platform, KMS and digital libraries to meet the learning needs, knowledge transfer requirements and capacity-building commitments of the IntelCities CoP. This, it has been suggested, marks a significant step forward in the development of learning services and offers the opportunity for platforms of this type to develop as a KMS supported by digital libraries. In view of this, the chapter has suggested that if the full significance of these technical innovations is to be realised, then this integration needs to progress and requires the e-learning platform, KMS and digital library developed for such purposes to not only be interoperable across the IntelCities middleware, but all the eGov services that are available to citizens at the front end. The way in which the IntelCities CoP proposes to achieve this is particularly innovative because the organisation offers a strategy to consolidate the underlying learning aspirations of city portals, but as particular types of eGov services that have previously remained beyond the reach of the platforms developed for such purposes. That is to say, out with the grasp of previous attempts which have been made by such organisations to develop a knowledge base capable of delivering the consultation and deliberation services that are key to all this.

Notes

1 An earlier version of this chapter was published as: Deakin, M., Lombardi, P. and Cooper, I., The IntelCities CoP for the Capacity-building, Co-design, Monitoring and Evaluation of eGov Services, Journal of Urban Technology, 2011, 18 (2): 17–38.
2 See http://www.intelcitiesproject.com.
3 The curious subtitle of the article by Amin and Roberts (2008) is 'beyond communities of practice': they use the phrase to suggest the need to 'get beyond' the 'undifferentiated' use of the term and for more contextualised studies of the type set out in this paper.

References

Amin, A. and Roberts, J. (2008) Knowing in Action: Beyond Communities of Practice, *Research Policy*, 37: 353–369.
Berger, C., Möslein, K., Piller, F. and Reichwald, R. (2005) Cooperation Between Manufacturers, Retailers, and Customers for User Co-design: Learning from Exploratory Research, *European Management Review*, 1: 70–87.
Binder, T., Brandt, E. and Gregory, J. (2008) Design Participation(s): A Creative Commons for Ongoing Change, *CoDesign*, 4 (2): 79–83.
Campbell, F. and Deakin, M. (2005) Cities as Leading Examples of Digitally-Inclusive Knowledge Societies: The e-Citizenship Course, Representative Users, Pedagogy and Engagement Matrix, in Osborne, M. and Wilson, B. eds, *Making Knowledge Work*, Stirling, Stirling University.
Curwell, C., Deakin, M., Cooper, I., Paskaleva-Shapira, K., Ravetz, J. and Babicki, D. (2005) Citizens' Expectations of Information Cities: Implications for Urban Planning and Design, *Building Research and Information*, 22 (1): 55–66.
Deakin, M. (2007) e-Topia, SUD and ICTs: Taking the Digitally-inclusive Urban Regeneration Thesis Full Circle, *Journal of Urban Technology*, 14 (3): 131–139.

Deakin, M. and Allwinkle, S. (2006) The IntelCities Community of Practice: The e-Learning Platform, Knowledge Management System and Digital Library for Semantically-Interoperable e-Governance Services, *International Journal of Knowledge, Culture and Change Management*, 6 (3):155–162.

Deakin, M. and Allwinkle, S. (2007a) e-Topia, SUD and ICTs: The Post-human Nature, Embedded Intelligence, Cyborg-self and Agency of Digitally-inclusive Regeneration Platforms, *International Journal of the Humanities*, 5 (2): 199–208.

Deakin, M. and Allwinkle, S. (2007b) Urban Regeneration and Sustainable Communities: The Role of Networks, Innovation and Creativity in Building Successful Partnerships, *Journal of Urban Technology*, 14 (1): 77–91.

Deakin, M., Van Isacker, K. and Wong, A. (2004) *Review of the IntelCities Knowledge Capture Requirements Using a S.W.O.T. Analysis*, Edinburgh, Edinburgh Napier University.

Ellis, D., Oldridge, R. and Vasconcelos, A. (2004) Community and Virtual Community, *Annual Review of Information Sciences and Technology*, 38: 146–186.

Graham, S. and Marvin, S. (1996) *Telecommunications and the City*, London, Routledge.

Johnson, J. (2001) A Survey of Current Research on Online Communities of Practice, *Internet and Higher Education*, 4: 45–60.

Josefsson, U. (2005) Coping with Illness Online: The Case of Patients' Online Communities, *The Information Society*, 21: 143–153.

Lombardi, P. and Curwell, S. (2005) INTELCITY Scenarios for the City of the Future, in D. Miller and D. Patassini eds, *Beyond Benefit Cost Analysis*, Aldershot, Ashgate.

Lombardi, P., Cooper, I., Paskaleva, K. and Deakin, M. (2009) The Challenge of Designing User-centric e-Services: European Dimensions, in Riddeck, C., ed., *Research Strategies for e-Government Service Adoption*, Hershey, Idea Group Publishing.

Mitchell, W. (2000) *e-Topia: Urban Life, Jim But Not as You Know It*, Cambridge, Massachusetts, MIT Press.

Nikolaus, F., Keinz, P. and Schreier, M. (2008) Complementing Mass Customization Toolkits with User Communities: How Peer Input Improves Customer Self-Design, *Journal of Product Innovation Management*, 25 (6): 546–559.

Orr, J. (1996) *Talking About Machines: An Ethnography of a Modern Job*, New York, IRL Press.

Osterlund, C. (1996) *Learning Across Contexts*, Arhus, Arhus University.

Wenger, E. (1998) *Communities of Practice: Learning, Meaning, and Identity*, Cambridge, Cambridge University Press.

Wenger, E. (2000) Communities of Practice and Social Learning Systems, *Organization*, 7 (2): 225–246.

Wind, J. and Rangaswamy, A. (2001) Customerization: The Next Revolution in Mass Customization, *Journal of Interactive Marketing*, 15 (1): 13–32.

5   What makes cities intelligent?[1]

Nicos Komninos

Introduction

The new interdisciplinary paradigm of 'intelligent cities' or 'smart cities', bringing together theories, methodologies and practices from many diverse fields, such as urban development and planning, information and communication technologies, web and internet technologies, and knowledge and innovation management, is overturning established urban development and planning. However, the impact of this paradigm reaches far beyond the domain of cities, as it influences the challenges of global competitiveness, sustainability and climate change, inclusion and employment.

A very rich literature reflects the evolution of thinking and practice in the field of digitally – intelligent – smart cities and outlines the contribution of information technologies and the internet to city development and planning, along with the role of knowledge processes and innovation in twenty-first-century cities. From Mitchell (1996), Ishida and Isbister (2000) and Graham (2003) focusing on technologies, experiences and case studies of digital cities, to Komninos (2002 and 2008) and Bell et al. (2009) on intelligent cities and the nexus of ICTs, broadband networks, collective intelligence and innovation that converge to sustain the development of cities, to Caragliu et al. (2009), Belissent (2010), Deakin (2011) and Schaffers et al. (2011) on smart cities, mobile devices and sensor networks embedded into the physical space of cities, this literature also highlights a trajectory of change. It describes the continuous evolution of digital technologies and innovation systems which feed intelligent cities and the creation of more open and innovative urban ecosystems deployed over the digital, social and physical space of cities. Such ecosystems enable citizens, end-users, enterprises and organisations to develop innovative attitudes and become more resource efficient and intelligent in decision making.

Despite the great diversity of strategies and solutions that can be observed, intelligent and smart cities rely on a core of knowledge processes. We call this core the 'spatial intelligence of cities'. This concept allows us to bring 'intelligent cities' and 'smart cities' into a common field of reference and makes it possible to focus on their fundamental processes. Spatial intelligence is made by informational, cognitive and innovation processes, such as information collection and processing, real-time alert, forecasting, learning, collective intelligence, distributed problem solving, collaborative innovation and co-creation, which take birth within cities and enable citizens and organisations to address more efficiently the challenges they face. It refers to the ability of a community to combine its intellectual resources, institutions for collaboration and smart infrastructure within knowledge networks that deal with a wide range of problems and challenges.

Spatial intelligence is the ingredient that makes cities intelligent. As the title suggests, the aim of this chapter is to discuss the different architectures and trajectories that were followed to make spatial intelligence emerge. Furthermore, to describe the fundamental variables of spatial intelligence and how they change alongside the evolution that takes place in digital technologies and innovation ecosystems.

The chapter is structured as follows. We start from the concept of spatial intelligence of cities and a quick overview of the literature on cyber, digital, intelligent and smart cities which outlines different types of spatial intelligence. Then we describe three different trajectories and architectures of spatial intelligence (orchestration, empowerment, instrumentation) which can be found within cities. At Bletchley Park large-scale, organised cooperation took place during World War II involving creative people, collective working procedures, rule-based thinking and the first use of intelligent machines; in Cyberport Hong Kong we find a set of urban infrastructures which continually improve the skills and talents of the Hong Kong younger population; and in Amsterdam Smart City we find smart systems for collecting and disseminating information in real time which enable citizens to take informed decisions and improve the environment of the city. In the last section, we come back to an argument presented in previous publications (Komninos, 2006 and 2008), namely, that the intelligence of cities is based on the integration of different city layers (physical, institutional, digital) and the activation of knowledge functions, which are collectively created and deployed, such as network-based information intelligence and forecasting, technology learning and acquisition, collaborative innovation, information dissemination and product promotion.

Spatial intelligence of cities

The literature on digital, intelligent and smart cities, which spans a period of twenty years, highlights different forms of spatial intelligence which appeared with respect to different web technologies, knowledge and innovation processes and forms of citizens involvement. The first academic paper on intelligent cities appeared in 1992 (Laterasse, 1992) and the first academic paper on smart cities was also published the same year (Gibson et al., 1992). Since then, these ideas have spread widely, both in theory and practice, and these literatures profoundly changed the dominant discussion of the 1980s and 1990s about cities, post-Fordism, production flexibility, technopoles and technology districts.

Two major forces have driven the paradigm shift towards intelligent cities. On the one hand, the rising knowledge and innovation economy that fuels contemporary economic development worldwide, and on the other, the spread of the internet and the web as a major technological innovation. The paradigm of intelligent cities brings these two trajectories together. Urban development has become linked with innovation ecosystems, knowledge-driven localities, innovation clusters and creative hubs, in which R&D, knowledge, innovation, people's creativity, learning and training are connected by forces of agglomeration and locality, trust, knowledge spillovers and tacit knowledge exchange. In parallel, a new digital spatiality has been added to the physical and institutional space of cities. It is composed of broadband networks, web applications and e-services, creating an umbrella of digital communication and cooperation which is situated over the cities. ICTs, the internet and the web alone would not have had a strong impact on cities if contemporary urban agglomerations had not rooted their development in knowledge and innovation.

The new digital spatiality has joined the spatialities of agglomeration and regulation in multiple ways, enhancing communication, city representation, virtualisation of infrastructures, changing activities, optimisation of city functions and governance. These different roles of the digital and the different forms of integration between physical, institutional and digital spaces have given birth to a series of concepts within the same event horizon, namely those of cyber, digital, intelligent and smart cities.

The cyber literature marked the initial stage of the digital spatiality of cities. Cybercities and cyberspace refer to any type of virtual space generated by a collection of data within the internet (Shiode, 1997), but the concept also contains the sense of inspection and control with communication and information feedback as preconditions of effective action. It carried some seeds from the ideas of cybernetics that appeared in the 1940s on communication with machines and feedbacks in decision making. This perspective led to early e-government applications for city management and recently to technologies for security and control over the urban space, and in some cases the transfer of military methods of tracking, identification and targeting into the governance of urban civil society (Graham, 2010). In a broader sense, a cybercity is conceived as a web-based city in which people interact with each other through and exclusively over cyber space. Anttiroiko (2005) notes that the 'cyber' prefix refers also to a dark side of the virtual space, its 'cyberterrorism' and 'cyborg' dimensions.

A more neutral discussion developed within the digital city literature and the extensive work of Ishida and Isbister (2000), Hiramatsu and Ishida (2001) and Van den Besselaar and Koizumi (2005). It concerned the representation of the city, in early forms via portal-type web pages, panoramic and 3D representations of cities, and later with augmented reality technologies and urban tagging. Digital cities are connected communities which combine 'broad-band communications infrastructure; a flexible, service-oriented computing infrastructure based on open industry standards; and innovative services to meet the needs of governments and their employees, citizens and businesses' (Yovanof and Hazapis, 2009). Here the digital city offers a metaphor of the city; an understanding of the city through its virtual representation. Such digital cities were described as 'mirror-city metaphors', as their logic was to offer 'a comprehensive, web-based representation, or reproduction, of several aspects or functions of a specific real city, open to non-experts' (Couclelis, 2004). The spatial intelligence of cities related to this type of solution was based on advantages of representation and visualisation. 'One picture is worth a thousand words' reflects this idea that complex environments can be described and understood better by a virtual representation or metaphor.

City intelligence comes to the scene with the understanding that digital spaces improve urban ecosystems, because of their capacity to process information and sustain learning, innovation and distributed problem-solving. Intelligent cities emerge at the crossing between the knowledge-based development of cities (knowledge cities) and the digital cities of media. According to Mitchell (2007) the intelligence of cities 'resides in the increasingly effective combination of digital telecommunication networks (the nerves), ubiquitously embedded intelligence (the brains), sensors and tags (the sensory organs), and software (the knowledge and cognitive competence)'. City intelligence comes from partnerships and social capital in organising the development of technologies, skills and learning, and engaging citizens to become involved in creative community participation (Deakin and Allwinkle, 2007). The intelligence of cities is based on a combination of the creative capabilities of the population, knowledge-sharing institutions and digital applications organising collective intelligence, which together increase the ability to innovate, and in turn offer the ultimate measure of intelligence. Therefore, it emerges within urban agglomerations from the integration of three types of intelligence: (1) the inventiveness, creativity and human intelligence of the city's population, (2) the collective intelligence of the city's institutions and social capital for innovation, and (3) the artificial intelligence of public and city-wide smart infrastructure, virtual environments and intelligent agents (Komninos 2008). From this perspective, the spatial intelligence of cities builds on collective intelligence and social capital for collaboration. It is based also on people-driven innovation introducing the principles of openness, realism and empowerment of users in the development of new solutions (Bergvall-Kåreborn and Ståhlbröst, 2009).

The recent turn and interest towards smart cities highlights two new concerns: on the one hand the pursuit of sustainability: a smart city should support a more inclusive, diverse and sustainable urban environment, green cities with less energy consumption and CO_2 emissions (Eurocities, 2009; Caragliu et al., 2009), and on the other hand, the rise of new internet technologies promoting real-world user interfaces with mobile phones, smart devices, sensors,

RFIDs, the semantic web and the Internet of Things. Smart city literature focuses on the latest advancements in mobile and pervasive computing, wireless networks, middleware and agent technologies, as they become embedded into the physical spaces of cities and are fed with data all around the clock. Smart city applications – with the help of instrumentation and the interconnection of mobile devices and sensors which collect and analyse real-world data – improve the ability to forecast and manage urban flows and push city intelligence forward (Chen-Ritzo et al., 2009). Within this technology stack spatial intelligence moves out of applications and enters into the domain of data and what data means becomes part of data. Here data is provided just-in-time, and real-time data enables real-time response to be made.

Critical questions within this large landscape of practices and transformations concern the sources of the spatial intelligence of cities: the structures, mechanisms and architectures that sustain the problem-solving capability of cities. What makes a city intelligent or smart? Which type of spatial intelligence is activated within each district/sector of the city? Is it a spatial intelligence common to all districts or are different structuring forms activated within different city districts depending on their functional characteristics (manufacturing, commerce, education, recreation, etc.) and governance?

We discuss these questions with respect to three case studies of cities (Bletchley Park, Cyberport Hong Kong and Amsterdam Smart City) and three forms of spatial intelligence. These forms are *orchestration intelligence*, which is based on collaboration and distributed problem-solving within a community having full control over information and knowledge processes; *empowerment or amplification intelligence*, which is based on people's up-skilling provided by experimental facilities, open platforms and city infrastructure; and *instrumentation intelligence*, based on real-time information, comparative data analysis and predictive modelling for better decision-making across city districts. These trajectories of spatial intelligence can work in isolation or in coordination. They provide different levels of problem-solving capability, but they always rely on network structures and connections between the physical, institutional and digital space of cities.

Cities

From the moment they emerged, cities were based on advantages created by spatial proximity such as division of labour and collaboration, use of common infrastructure, face-to-face communication and the development of trust. The spatial agglomeration of people, activities, buildings and infrastructure was made possible by advances in the division of labour and exchange of goods, and in turn generated a series of positive social and economic externalities. Soja (2003), writing about the first urban settlements and cities, insists on 'putting cities first', attributing to synekism –the physical agglomeration of people with a form of political coordination – the capacity to advance creativity, innovation, territorial identity and societal development which arises from living in dense and heterogeneous agglomerations. Soja refers extensively to *The Economies of Cities* by Jane Jacobs (1969) and the findings in Catal Huyuk, the largest and most developed early city in southern Anatolia, where Jacobs located major innovations and transformations from hunting and gathering to agriculture, the first metallurgy, weaving and crude pottery, which took place because of the existence of the city. These innovations, he argues, as well as every major innovation in human society, come from cooperation, synergy and the multiple savings obtained from living in dense urban settlements. The creative externalities of cities – various types of agglomeration economies, external, scale, scope, location, urbanisation – stem, on the one hand, from savings in energy, time and materials, and on the other hand, from collaboration and the creation of synergies. The spatial agglomeration of people and activities produces both savings and synergies. New industrial geography has explained how proximity generates additional externalities in the innovation economy because of informal collaboration, untraded interdependences, knowledge spillovers, trust and the transmission of tacit knowledge.

With the digital spatiality extended over cities, collaboration and synergies are scaling up. As citizens come into the digital space they share more and share it quicker. Interaction becomes easier and synergy stronger. The holy triad of synergies – proximity, trust, communication – is strengthened: proximity increases because the 'other' is just a few clicks away; trust deepens because digital interaction leaves traces; communication intensifies because we have more means and tools for interaction. Digital interaction enables wider collaboration, more extended supply chains and more end-user participation. Multiple digital technologies enhance the scalability of collaboration, such as co-design tools, collaborative work environments, real-time communication without cost, crowdsourcing solutions and content mash-ups.

As computers, devices and information systems become embedded into cities, the collaboration patterns among citizens change substantially. Change does not concern scalability only, but the architectures of cooperation as well. New networking architectures emerge, involving both humans and machines. As digital technology transfers tasks from humans to machines, workflows become more complex, more tasks are performed by cooperation, machines inspect the workflow of collaboration and storing capacity skyrockets. The city ends up with higher problem-solving capability, quicker responses, better quality procedures and lower operation costs; in other words, with higher spatial intelligence. This happens because machine intelligence is added to the human intelligence of citizens and to the collective intelligence of their community.

Orchestration: Bletchley Park, the first intelligent community

The first community that successfully practised this type of human–machine cooperation and integration of individual, collective and machine intelligence was Bletchley Park in the UK. The story of Bletchley Park is well known in the World War II code-breaking literature. However, it was never referred to as an intelligent city or intelligent community.

Bletchley Park is located 80 kilometres north-west of London. Bletchley is an ordinary town, a regional urban centre in the county of Buckinghamshire, at the intersection of the London and North-Western Railway with a line linking Oxford to Cambridge. Just off the junction, within walking distance from the station, lies Bletchley Park, an estate of about 100 hectares with a grand Victorian mansion at the centre of the estate.

The development of Bletchley Park started in August 1939, when the Government Code and Cypher School moved from London to Bletchley Park to carry out their code-breaking work in a safer environment. A small group of people initially settled at Bletchley, composed of code-breaking experts, cryptanalytic personnel and university professors from the exact sciences and mathematics. Alan Turing arrived at Bletchley Park in 1939 together with other professors from Cambridge to help set up the methods of analysis and workflow. Bletchley Park carries the mark of Turing and his ideas on intelligence, logic and software priority over hardware, and solutions over a universal computing process. The work was done in wooden huts, designated by numbers, and brick-built blocks which were constructed after 1939 to house the different sectors of cryptanalysis. In the years thereafter, the personnel of Bletchley Park increased in number at a spectacular rate and by the end of the war they numbered about ten thousand. People came from all fighting services, seconded to Bletchley Park because of their skills, and included civilians, authors, diplomats, bankers, journalists, teachers and many women who received training in information processing tasks.

The mission of Bletchley Park was to find the daily settings of the Enigma machines used by the German army to encode all transmitted messages between the army headquarters, divisions, warships, submarines, port and railway stations, military installations and other installations, and then decode all these messages. It is estimated that by 1942 the German army had a least a hundred thousand Enigma machines, which produced an enormous traffic of codified messages of vital importance for the daily operation of all army units. The Enigma machine was an electro-mechanical device for the encryption and decryption of messages based on polyalphabetic substitution. It relied on interchangeable rotors of 26 letters, initially three and later five moving rings and a plugboard which permitted variable electrical wiring connecting letters in pairs. Every key press on the keyboard caused one step on the first rotor – after a full rotation the others rotors also moved – and then electrical connections were made that changed the substitution alphabet used for encryption. Decoding was symmetrical. The receiver had to settle the machine in its initial setting of rotors, rings and plugging, type the coded message and recover the original. The combination of rotor order, the initial position of rotors and plug settings created a very large number of possible configurations. For each setting of rotors there were trillions of ways to connect ten pairs of letters on the plugboard. It was practically impossible to break the encryption by hand.

The amount of collaborative knowledge work carried out at Bletchley Park was enormous. The park was an 'industry' for information collection, processing, decoding and distribution. About 2,000 to 6,000 messages were processed and translated daily, while overall 200,000–500,000 German messages were decoded between 1940 and 1945. The impact was also extremely high. The strategic role of Bletchley Park was in the battle for supplies, defeating the U-boats in the Atlantic and securing the inflow of materials, foods and ammunition to Britain. By the end of 1941 the British announced that the problem of maritime supplies had been solved. Historians estimate that the work done in Bletchley Park shortened the war by two to four years and saved millions of lives. The philosopher George Steiner described Bletchley Park as the greatest achievement of Britain during the war and perhaps during the whole of the twentieth century.

The work done at Bletchley Park in breaking German communications codes was based on a collaborative workflow between scientists, experts, trained workers and machines which offered increased intelligence to deal with this challenge. The system had all the four essential characteristics that we now attribute to intelligent cities: (1) a creative population working in information and knowledge-intensive activities; (2) institutions and routines for collaboration in knowledge creation and sharing; (3) technological infrastructure for communication, data processing and information analysis; and (4) a proven ability to innovate and solve problems that appear for the first time. Bletchley Park was the first intelligent community ever created.

The methodological solutions about how to break the Enigma ciphers were given by a group of British cryptanalysts and mathematicians at Bletchley Park who continued and enriched the methods devised by Polish mathematicians in previous and simpler models of Enigma machines. The wiring structure of the machines and some fundamental design flaws – no letter could ever be encrypted as itself – were exploited. The breaking of the codes was based on human factors and mistakes made by the Germans. Alan Turing and Cambridge mathematician Gordon Welchman, who also invented the method of perforated sheets, provided the designs for the new machine – the British Bombe – which could break any Enigma cipher, provided there was an accurate assumption of about twenty letters in the message. Alan Turing contributed with several insights in breaking the Enigma, somehow continuing his theoretical work on computable numbers and the Turing universal machine.

Key to the success of Bletchley Park was collaboration and workflow integrating the whole information analysis process. Cryptanalysts worked as a team. They had to analyse all the messages of the day to make assumptions from the basic setting of the rotors. Codebooks found in sunken submarines or captured ships were also very helpful and provided Enigma ground settings and abbreviations. They had to simulate the entire German classification system, mapping and acronyms. Cryptanalysis acquired meaning only through the coordination of different activities across an extended workflow, and solving ciphers was only part of it. There was organised division of labour and specialisation into different tasks along the process of intercepting the messages, transferring them to Bletchley Park, code breaking, verification and dissemination to recipients of the information. The raw material came from a web of wireless intercept stations around Britain and overseas. Code-breakers based in the huts were supported by teams who turned the deciphered messages into intelligence reports. The letter from Turing, Welchman, Alexander and Milner-Barry to Churchill in October 1941, asking for more resources at Bletchley Park, personnel, night shifts, interception stations, specialised decoders and support to the Bombes, shows this integrated functioning of the community.

When a cryptanalyst developed an assumption about a possible way of breaking the code in a message, he prepared a menu (called a crib – plain text that corresponded to the cipher text) which was sent to be tested on a Bombe machine. This was an electromechanical machine used to discover the set of rotors, the settings of the alphabet rings and the wiring of the plugboard. The machine would check a million permutations, exclude those containing contradictions, and finally reveal how the Enigma machine had been set in order to produce this crib. The Bombe would then provide a solution by discounting every incorrect one in turn. The first Bombe was based on Turing's design and was installed at Bletchley Park in 1940. Subsequent Bombes were equipped with Welchman's diagonal board which could substantially decrease the number of possible rotor settings. In 1944 Colossus, the first digital electronic computer, became operational at Bletchley Park. Colossus was designed to break messages coded on Lorenz machines. The Lorenz machine created more complex ciphers using a code in which each letter of the alphabet was represented by a series of five electrical impulses. Obscuring letters were also generated by Lorenz's 12 rotors. The first Colossus arrived at Bletchley Park in December 1943 and in practical terms Bletchley Park used the world's first electronic computer and digital information processing machine.

Bletchley Park was a prototype of an intelligent community, an urban ecosystem in which the division of labour and orchestration of distributed tasks, workflows based on institutional rules and intelligent machines produced radical innovations. The military organisation in this district and absence of the spontaneous complexity we find in cities should not undervalue the innovativeness of its design and its effectiveness in dealing with extremely complex problems. Besides all merits, it represents a top-down solution feasible under extreme conditions when the social division of labour within cities also becomes a technical division.

Empowerment: CyberPort Hong Kong up-skilling infrastructure

There are, however, other routes to the spatial intelligence of cities, which leverage the impact of knowledge-intensive infrastructures and districts that are currently shaping the city's built environment.

The spatial structure of knowledge-based and intelligent cities is actually taking the form of 'knowledge ecosystems and city districts over smart networks'. This form is partly due to the need for active management of technological infrastructure and innovation ecosystems, and partly to the development of smart urban networks. The literature on the clustering of innovation has explained the causes of spatial agglomeration and the creation of islands of innovation (Morgan, 2004; Simmie, 1998). Many types of clusters – cohesion, industrial districts, innovative milieu, proximity – with different degrees of internal association and input–output interaction (Hart, 2000) and different sizes conglomerate over the city infrastructure. City networks for mobility, energy, water and utilities, on the other hand, are becoming smarter under the pressure of environmental sustainability and the need to save resources. It is estimated that smart infrastructure, smart grids, sensors, wireless meters and actuators might have a higher impact on energy saving and CO_2 reduction than the total positive effect from renewable energy sources.

Metropolitan plans like the Melbourne 2030 Plan and Stockholm's Vision 2030 have clearly adopted this strategy of organising various types of innovation ecosystems and knowledge-intensive districts over advanced infrastructure, including broadband, telecommunications, energy, multimodal transport and logistics to sustain the development of the innovation economy. Melbourne has institutionalised this type of development via 'knowledge precincts', areas surrounding university campuses in which special land use regulations favour the location of activities that link university infrastructure and R&D, offering opportunities for technology diffusion and cross-fertilisation between high-tech businesses, academia and public sector facilities (Yigitcanlar et al., 2008).

All innovation ecosystems profit from technology networking, knowledge spillovers and knowledge transfer. However, some ecosystems are pursuing conscious strategies for involving the wider population of the city, not just producers and technologists, and are creating a flow of up-skilling with education and learning on experimental facilities and ICT infrastructure. In the case of Living Labs, for instance, users are involved in new product development and testing within real urban environments. Participatory innovation processes integrate *co-creation* activities, bringing together technology push and application pull, *exploration* activities engaging user communities in an earlier stage of the co-creation process, *experimentation* activities, implementing the proper level of technological artefacts to experience live scenarios, and *evaluation* of new ideas and innovative concepts as well as related technological artefacts in real life situations (Pallot, 2009). To date, after six successive waves of expansion, the European Network of Living Labs (ENoLL) has more than 300 members from the 27 EU member states and countries outside the EU, such as Switzerland, Canada, the USA, China, Taiwan, Brazil, Mozambique, Senegal and South Africa. These open and user-centric innovation ecosystems operate in many and diverse activity sectors, such as mobile communications, media, agriculture, food industry, health, medicine, e-government services, smart cities, sports, education and social work.

There are also city ecosystems which act as 'innovation universities' or 'intelligent campuses', which use the built environment of the city and experimental facilities to involve citizens in learning and innovation. Large-scale up-skilling strategies thus become possible, thereby improving the creativity, intelligence and inventiveness of the population, and introducing an 'innovation for all' environment, in which every citizen can become a producer of services and innovations.

Cyberport Hong Kong is an innovation ecosystem that has effectively advanced this strategy of up-skilling, using advanced telecommunication infrastructure and multimedia technologies. It is a new knowledge district located on the west side of Aberdeen Country Park on Hong Kong island. The district has been developed as a government project aimed at developing the knowledge economy throughout Hong Kong. As an independent technology district, Cyberport is focusing on professional and enterprise development, offering an open platform for creative ideas to flourish and start-ups to be created in the field of media technologies. The district is wholly owned by the Hong Kong SAR government and managed by the Cyberport Management Company Limited.

Cyberport includes many different activities and land uses. Within a relatively small piece of land of 24 hectares there is an enterprise zone with four quality buildings which host about 100 information technology and media companies, a research institute, business incubator, conference centre, shopping mall, five-star hotel (Le Meridien), a huge housing complex and a large park at the heart of Cyberport, which also extends along the coastline. The area is served by fibre optic and copper networks offering high-speed broadband connections and a wide range of digital services and laboratory equipment. Buildings in the technology zone are grade A intelligent office buildings. All these activities are organised into four different zones: the technology zone with Cyberport 1, 2, 3 and 4 buildings; the commercial zone with the mall and the hotel; the residential zone; and the park and open area zone. Despite this functional division, the relatively small surface of the district and the openness here create a continuum of uses, as all the spaces are accessible to the community of the district.

Activities and land uses have been selected to promote the mission of the district and ensure its sustainability. Cyberport was developed on public land and the construction work took place from 2000 to 2008. The funding scheme foresaw a split into two parts, the Cyberport zone and the ancillary residential zone. The mission of the Cyberport zone was to create a strategic cluster of leading information technology and information services companies and a critical mass of professionals in these sectors. The mission of the residential zone was to generate revenue for the Cyberport project. A development company acquired part of the land (about 20 per cent of the plot) together with the infrastructure already on site to build the residential zone. The developer

(Cyber-Port Limited) was responsible for the total construction costs of both the Cyberport and the housing complex (Hong Kong Legislative Council, 2002). The residential zone includes eight 50-storey high buildings and two lower complexes – two to five storeys – for high-income residences along the coast. Overall 2,800 homes were built. In return for the concession on the land and infrastructure of the residential zone, the developer delivered the technology zone as a turn-key solution, with Cyberport 1, 2, 3 and 4, the shopping mall arcade and the five-star hotel operated by Le Meridien. Revenue generated from the commercial zone – mall and hotel – flows into the technology zone and covers the training, learning and incubation expenses. The district was publicly funded and serves the public interest. This genuine funding model contributes both to the development and operation of Cyberport 1, 2, 3 and 4, and to the public and open character of the district.

Cyberport should not be seen as a normal technology district or technology park. It is an ecosystem that nurtures talent in the media industry, turning skills and talent into start-ups. It amplifies the skills and creativities of the Hong Kong population using experimental digital infrastructure and open platforms. The objectives are technology diffusion, up-skilling and the enhancement of human capabilities. Cyberport is a creative community supplied with advanced communication and media infrastructure and digital connectivity.

> Cyberport identifies, nurtures, attracts and sustains talent so it is able to mobilize ideas, talents and creative organisations. It is a creative milieu; a place that contains the necessary requirements in terms of hard and soft infrastructure to generate a flow of ideas and inventions. (interview with CEO of Cyberport, N. Yang)

The focus of the district is the IT and multimedia sector, where it sustains a creative community. Technologies and applications that have been developed in Hong Kong universities or the Technology Park can be transferred to the younger generation though practical learning and experimental training. Training from the world's leading media and IT companies is provided together with the laboratory equipment and start-up funding for follow-up training which promotes entrepreneurship.

To achieve these objectives Cyberport developed state-of-the-art infrastructure, media equipment and digital services which are organised as open technology platforms. Each platform serves a specific objective of training, creativity and entrepreneurship:

- *The Digital Entertainment Incubation and Training Programme* is a platform whose objective is to build and promote entrepreneurship and competence in the digital entertainment industry, focusing on business skills, games, animation and digital entertainment, and to enhance networking with industry, as well as to promote the awareness and interest of the younger generation in digital entertainment.

- *The Digital Media Centre* is a unique state-of-the-art digital multimedia creation facility, whose objective is to offer software and hardware support to content developers, multimedia professionals and small- and medium-sized enterprises.
- *The iResource Centre* is a digital content storage platform, which serves as a trusted marketplace and clearing house for the aggregation, protection, licence issuance and distribution of digital content.
- *The Testing and Certification of Wireless Communication Platform* provides continuous mobile communication services and coverage of mobile phone signals (3G, GSM, CDMA and PCS) in both outdoor and indoor areas within Cyberport in cooperation with major mobile communications service operators.
- *The Cyberport Institute* was established by the University of Hong Kong to introduce and run IT courses for talented people and to support various IT development and related businesses in Hong Kong.

These open technology platforms are operated in cooperation with industry leaders who are the founding industrial partners. Cisco, Hewlett Packard, IBM, Microsoft, Oracle and PCCW have been involved through sponsorship programmes, while the students benefit from access to top-of-the-market technologies, scholarships, placement opportunities and employment.

The dual mechanism described above – the open digital technology platforms and real-estate based sustainability – provides an open-ended mechanism for professional training and up-skilling. The setting enhances human capabilities and intelligence by simultaneously providing hard urban infrastructure and soft digital technologies and services. Developed on public land, Cyberport is creating intelligence through real-estate business models, skills and human development programmes which spreads out into the entire urban system of Hong Kong.

Instrumentation: Amsterdam Smart City real-time decision making

One of the most significant recent contributions to the intelligence of cities debate comes from the initiative developed by IBM, Smart Planet – Smarter Cities. IBM proposes a city intelligence solution based on the combination of networks, smart meters and data modelling. These technologies can optimise the use of infrastructure and the finite city resources, driving efficiency and effectiveness, by making city systems: (1) interconnected, (2) instrumented and (3) intelligent. *Interconnection* means that different parts of a core system can be joined and communicate with each other, turning data into information. *Instrumentation* of a city's system means that running that system produces data on key performance indicators and the system becomes measurable with instruments and smart meters. *Intelligence* refers to the ability to use the information gathered to model patterns of behaviour and develop predictive models of likely outcomes, allowing better decision making and informed actions (Dirks and Keeling, 2009; IBM, 2010). It is estimated that such instrumented intelligence might cut city traffic by as much as 20 per cent, save energy by up to 15 per cent, lower the cost of therapy by as much as 90 per cent and substantially reduce the city's budget spent on public safety (Kaiserswerth, 2010). IBM is testing this concept through partnerships with cities worldwide. In many cities the company and local administrations work together to provide services in energy, water management and transportation, reducing the city's impact on the environment. Pilot testing and experimental facilities provide information on how to consume better electricity, water, natural gas and oil.

The same concept is pursued at research level from the Future Internet Research (FIRE) initiative, the European FP7-ICT programme that is funding the experimental facility of Smart Santander in the city of Santander, in northern Spain. The facility has developed a network of 12,000 sensors and devices monitoring pollution, noise, traffic and parking. The testbed is composed of around 3,000 IEEE 802.15.4 devices, 200 GPRS modules and 2000 joint RFID tag/QR code labels deployed both at static locations (streetlights, public buildings, facades, bus stops) as well as on board mobile vehicles (buses and taxis). Devices work over a common IP infrastructure using cellular, radio-meshed networks and available broadband (Krco, 2010). The facility is open to researchers and service providers to test architectures, enabling technologies and pilot applications, the interaction and management of protocols, and support services such as discovery, identity management and security, and the social acceptance of services related to the Internet of Things (see www.smartsantander.eu).

Instrumentation intelligence is also widely pursued in Amsterdam Smart City. Smart devices and wireless meters transmit information over broadband networks and provide intelligence that citizens and organisations of the city can use to optimise their practice. Decisions can be made with respect to accurate and on time information provided by smart devices or by the crowd. Many solutions for this type of logic are being implemented in different districts of the city: housing and living (West Orange, Geuzenveld, Haarlem, Onze Energie), working (ITO Tower, monumental buildings, employee contest), mobility (Ship to Grid, Moet je Watt) and public space (Climate Street, smart schools, ZonSpot, smart swimming) (Baron, 2011). Overall 30 projects are implemented in three areas (Ijburg, Nieuw West, Zuid Oost) and five themes (Living, Working, Mobility, Public Facilities, Open Data) (see http://amsterdamsmartcity.com/).

In the Haarlem area, for example, 250 users can test an energy management system and get insight into the energy consumption of appliances, enabling monitoring of energy usage and appliances to be remotely switched on and off. In the Geuzenveld neighbourhood, 500 homes have been provided with smart meters and energy displays to become aware of energy consumption and discuss energy savings at brainstorming sessions. In the West Orange project, 500 households have been provided with smart meters and displays and a personal energy saving goal is set for every household. The overall objective is to save at least 14 per cent of energy and reduce CO_2 emissions by an equal amount. The ITO tower, a large multi-tenant office building, is testing which smart building technologies, cooperative agreements and practices can make office buildings more sustainable. Information gained by smart plugs and insight based on data analysis will be used to provide more efficient solutions. In the Utrechtsestraat, a shopping street with numerous cafés and restaurants, 140 small enterprises are testing solutions for a more sustainable environment: logistics using electric vehicles, energy-saving lamps for street lighting dimmed during quiet times, solar-powered garbage compacters, smart meters and displays for energy consumption, and incentives and benefits arising from energy savings (Amsterdam Smart City, 2009). Amsterdam Smart City also recently experimented with crowdsourcing, co-creation and open innovation to involve citizens in finding better solutions for public space and mobility. Ambitious goals were set to reduce CO_2 emissions by 40 per cent and energy by 20 per cent in 2025 from the 1990 baseline. Key performance indicators show that these goals can be achieved. In the Climate street more than 50 per cent of sustainable waste collection and 10 per cent energy saving are already being recorded.

Towards a universal architecture of spatial intelligence

Orchestration, empowerment and instrumentation intelligence illustrate different architectures within the spatial intelligence of cities. They correspond to large-scale and city-wide manifestations of fundamentally different types of intelligence, namely collective (in some cases organisational), human and machine intelligence. All architectures increase the efficiency of cities to address complex and non-linear challenges, but they do this in very different ways. Orchestration stands on the large-scale division of work and the integration of knowledge tasks which are distributed among the members of a community. Each task may be simple, but the size of the collaboration defines the complexity of the entire knowledge process. The overall result may be truly innovative. Empowerment stands on improvements of individual skills and knowledge. It is an individual learning process, but if practised massively on the city level can produce great results. Instrumentation intelligence replicates computer processes at the city level, gathering data from sensors, social media and urban activities, and processing this information in real time.

A few variables, however, generate the above types of spatial intelligence:

- the space where knowledge processes take place: physical, institutional, digital space;
- the knowledge generated: information-in, learning, innovation, information-out;
- the intelligence involved in knowledge generation: human, collective, machine.

Clearly, orchestration, empowerment and instrumentation are not the only feasible forms of spatial intelligence produced from these variables. Evidently, many more combinations are possible. For instance, some cities offer other types of empowerment, such as digital spaces and crowdsourcing platforms for co-creation and innovation. Meanwhile, Future Internet and future media research are bringing in new solutions in terms of infrastructure (cloud computing, RFIDs, sensors, real world user interfaces, mobile devices), data (open data, linked data) and trusted services. However, the variables of spatial intelligence are technology agnostic.

Intelligent/smart city solutions are now being engineered in all domains of cities: the innovation economy of cities with the different city districts, sectors of economic activity, clusters and ecosystems that they contain; the quality of life with e-services for social care, health, safety, environmental monitoring and alert; the utilities of cities with their different networks, flows and infrastructures; and the governance of cities with services to citizens, decision-making procedures, participation and more direct democracy. At least 25 different domains of cities can be identified as potential fields of intelligent/smart city solutions with thousands of applications and e-services.

In each of these domains (X district or sector) spatial intelligence emerges from a combination of knowledge processes, type of intelligence involved and space of their deployment (Figure 5.1). Outputs and effectiveness in terms of city growth, employment and environmental sustainability depend on these knowledge variables. This is a critical issue for smart cities governance. Instrumentation seems suitable for providing resource-efficient urban networks, transport and energy services and utilities; orchestration offers advantages of competitiveness to industry clusters, ports and technology districts; empowerment is a good solution for innovative service clusters, living labs, start-ups, employment and city governance.

Smart or intelligent cities are expected to contribute to urban challenges with sound solutions. However, to date most smart/intelligent city solutions have had limited impact on the competitiveness, employment and sustainability of cities. This mismatch signifies several things: either smart cities are not well targeted on city challenges, that solutions are more technology push than demand driven, or that cities have not efficiently developed spatial intelligence. All explanations can be true, and cities with all the technology and institutions they actually have are not yet sufficiently intelligent. By and large, contemporary solutions are lagging in terms of the level of achievement and social impact reached by Bletchley Park.

It seems that we still lack a deeper understanding about what makes a city intelligent. We are still in the age of the digital, rather than intelligent, or smart, cities. All definitions of intelligent/smart cities stress the use of information and communications technologies to make cities more innovative and efficient. But they do not equally stress the need to understand the drivers of intelligence and the forms of integration among innovation actors, open, connected communities, their service applications, monitoring and

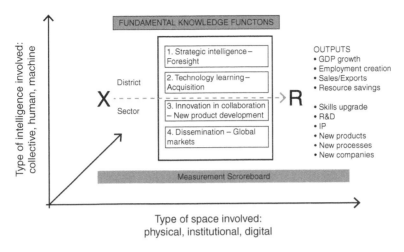

Figure 5.1 Generic dimensions of spatial intelligence of cities

measurement,which altogether change the knowledge and innovation functions of cities. We have to engineer integrated solutions for every sector, district and innovation ecosystem of a city, as integration is the key to higher spatial intelligence and efficiency.

Notes

1 An earlier version of this paper appeared as: Komninos, N., Intelligent Cities: Variable Geometries of Spatial Intelligence, *Journal of Intelligent Building International*, 3, 2011, 1–17.

References

Amsterdam Smart City (2009) *Introduction Amsterdam Smart City*, Municipality of Amsterdam.

Anttiroiko, A.V. (2005) Cybercity, *Encyclopedia of the City*, London, Routledge.

Baron, G. (2011) *Amsterdam Smart City*, Amsterdam Innovation Motor, unpublished paper.

Belissent, J. (2010) Getting Clever about Smart Cities: New Opportunities Require New Business Models, Forrester for Ventor Strategy Professionals.

Bell, R., Jung, J. and Zacharilla, L. (2009) *Broadband Economies: Creating the Community of the 21st Century*, Intelligent Community Forum Publications.

Bergvall-Kåreborn, B. and Ståhlbröst, A. (2009) Living Lab: An Open and Citizen-centric Approach for Innovation, *International Journal of Innovation and Regional Development*, 1 (4): 356–370.

Caragliu, A., Del Bo, C. and Nijkamp, P. (2009) Smart Cities in Europe, Research Memoranda 0048, VU University Amsterdam, Faculty of Economics, Business Administration and Econometrics.

Chen-Ritzo, C.H, Harrison, C., Paraszczak, J., and Parr, F. (2009) Instrumenting the Planet, *IBM Journal of Research* and *Development*, 53 (3): 338–353.

Couclelis, H. (2004) The Construction of the Digital City, *Environment and Planning B: Planning and Design*, 31: 5–19.

Deakin, M. (2011) The Embedded Intelligence of Smart Cities, *Intelligent Buildings International*, 3 (3): 189–197.

Deakin, M. and Allwinkle, S. (2007) Urban Regeneration and Sustainable Communities: The Role of Networks, Innovation and Creativity in Building Successful Partnerships, *Journal of Urban Technology*, 14 (1): 77–91.

Dirks, S. and Keeling, M. (2009) *A Vision of Smarter Cities*, Dublin, Ireland, Centre for Economic Development.

Eurocities (2009) *Smart Cities': Workshop Report*, 16–17 November 2010, Brussels.

Gibson, D.V., Kozmetsky, G. and Smilor, R.W., eds (1992) *The Technopolis Phenomenon: Smart Cities, Fast Systems, Global Networks*, New York, Rowman and Littlefield.

Graham, S., ed. (2003) *The Cybercities Reader*, London, Routledge.

Graham, S. (2010) *Cities under Siege: The New Military Urbanism*, London, Verso Books.

Hart, D.A. (2000) Innovation Clusters: Key Concepts, *Working paper, Department of Land Management and Development, and School of Planning Studies*, The University of Reading, United Kingdom.

Hiramatsu, K. and Ishida, T. (2001) An Augmented Web Space for Digital Cities, *Symposium on Applications and the Internet Proceedings*. Online, http://ieeexplore.ieee.org/xpl/freeabs_all.jsp?arnumber=905173.

Hong Kong Legislative Council (2002) Background Brief on Cyberport, Legislative Council Secretariat. Online www.legco.gov.hk/yr01–02/english/panels/itb/papers/itb0708cb1–2172–1e.pdf.

IBM (2010) *A Vision of Smarter Cities: How Cities Can Lead the Way into a Prosperous and Sustainable Future*, IBM Global Services.

Ishida T., and Isbister K., eds (2000) *Digital Cities: Technologies, Experiences, and Future Perspectives*, Berlin, Springer-Verlag.

Jacobs, J. (1969) *The Economy of Cities*, New York, Random House.

Kaiserswerth, M. (2010) *Creating a Smarter Planet: One Collaboration at a Time*, IBM Research Zurich, Online www.earto.eu/fileadmin/content/01_Seminars___Conferences/AC_2010/4-Matthias_Kaiserswerth.pdf.

Komninos N. (2002) *Intelligent Cities: Innovation, Knowledge Systems, and Digital Spaces*, London, Routledge.

Komninos, N. (2006) The Architecture of Intelligent Cities, *Intelligent Environments 06*, Proceedings, Institution of Engineering and Technology, 13–20.

Komninos N. (2008) *Intelligent Cities and Globalisation of Innovation Networks*, London and New York, Routledge.

Krco, S. (2010) *SmartSantander: A Smart City Example, ICT Event 2010*, Brussels, Belgium, 27–29 September.

Laterasse, J. (1992) *The Intelligent City*, in F. Rowe and P. Veltz, eds, *Telecom, Companies Territories,* , Paris, Presses de L'ENPC.

Mitchell, W. (1996) *City of Bits: Space, Place and the Infobahn*, Cambridge, MA, Massachusetts Institute of Technology.

Mitchell, W. (2007) Intelligent Cities, *e-Journal on the Knowledge Society*, online www.uoc.edu/uocpapers/eng.

Morgan, K. (2004) The Exaggerated Death of Geography: Learning, Proximity and Territorial Innovation Systems, *Journal of Economic Geography*, 4 (1): 3–21.

Pallot, M. (2009) Engaging Users into Research and Innovation: The Living Lab Approach as a User Centred Open Innovation Ecosystem, online www.cwe-projects. eu/bscw/bscw.cgi/1760838?id=715404_1760838.

Schaffers, H., Komninos, N., Pallot, M., Trousse, B., Nilsson M. and Oliveira, A. (2011) Smart Cities and the Future Internet: Towards Cooperation Frameworks for Open Innovation, The Future Internet, *Lecture Notes in Computer Science*, 6656: 431–446.

Shiode, N. (1997) An Outlook for Urban Planning in Cyberspace: Toward the Construction of Cyber Cities with the Application of Unique Characteristics of Cyberspace, *Online Planning Journal*, Centre for Advanced Spatial Analysis, University College London, online www.casa.ucl.ac.uk/planning/articles2/olp.htm.

Simmie, J.M. (1998) Reasons for the Development of 'Islands of Innovation': Evidence from Hertfordshire, *Urban Studies*, 35 (8): 1261–1289.

Soja, E. (2003) Writing the City Spatially, *City*, 7 (3): 269–280.

Van den Besselaar, P. and Koizumi, S. (2005) Digital Cities III. Information Technologies for Social Capital: Cross-cultural Perspectives, *Third International Digital Cities Workshop*, Amsterdam, Berlin, Springer-Verlag.

Yigitcanlar, T., O'Connor, K. and Westerman, C. (2008) The Making of Knowledge Cities: Melbourne's Knowledge-based Urban Development Experience', *Cities*, 25 (2): 63–72.

Yovanof, G.S. and Hazapis, G.N. (2009) An Architectural Framework and Enabling Wireless Technologies for Digital Cities and Intelligent Urban Environments, *Wireless Personal Communications*, 49 (3): 445–463.

6 The embedded intelligence of smart cities

Mark Deakin

Introduction

This chapter offers an extensive review of Mitchell's thesis on the transition from the city of bits (intelligent) to e-topia (smart cities). It suggests that the problems encountered with the thesis lie with the lack of substantive insight it offers into the embedded intelligence of smart cities. The chapter also suggests that if the difficulties experienced were only methodological they may perhaps be manageable, but the problem is they run deeper and relate to more substantive issues which surround the trajectory of the thesis. This is a critical insight of some significance because if the trajectory of e-topia is not in the direction of either the embedded intelligence of smart cities, or the ICTs of digitally inclusive regeneration platforms, then the question arises as to whether the thesis can be a progressive force for change.

Mitchell's thesis

Mitchell's (1995) book *City of Bits* sets out a vision of urban life literally done to bits, left fragmented and in danger of coming unstuck. Mitchell's (1999) next book on e-topia provides the counter-point to this vision of urban life, a scenario where the city is no longer left in bits and pieces, but a place where it all comes together. As Mitchell (2004) states in his more recent book, *Me++: the Cyborg Self and the Networked City*, all this 'coming together' is possible because: 'the trial separation of bits and atoms is now over' and this 'post-AD 2000 dissolution of the boundaries between the virtual and physical' is what makes everything worth playing for (p. 3). Worth playing for because this 'coming together' of the virtual and physical is something that not only needs to be networked, but embedded in the intelligence that architects, planners, engineers and surveyors require to make cities smart (ibid.).

While this thesis on the coming together of the virtual and physical and dissolution of the boundaries between cyber and meat space is compelling, it has to be recognised there are a number of concerns surrounding the technical, cultural, environmental and social status of the embedded intelligence currently available for planners, architects, engineers and surveyors to make cities smart.

Concerns with the status of Mitchell's thesis

The first rests with the ability of the thesis to cope with what Mitchell refers to as 'ancient concerns' surrounding the ecology and equity of urban development and the sustainability of the lean, mean and green strategy advanced to explain information society's process of dematerialisation (what he refers to as the shedding of atoms). Here the concern is not so much with the utopian legacy of such a vision, but the tendency the thesis has to 'repeat the mistakes of the past' by failing to acknowledge that techno-topian solutions of this kind leave cities without the means by which to deal with the ecology and equity of development. This is because for the likes of Graham and Marvin (1996, 2001) any such absence of means is seen as leaving the thesis open to the accusation of cultivating yet another kind of environmental determinism, albeit on this occasion one with a particularly keen interest in the networks assembled as the embedded intelligence of smart cities.

Mitchell's response to such accusations is tactical and astute, because while culturally locked into the environmental determinism of the techno-topian legacy drawn attention to by Graham and Marvin, the thesis manages to side-step this issue. Moreover, it manages to do this by asserting the late modern experience of urban life shall be neither utopian, nor dystopian, but e-topian. While a clever tactic, in the sense Mitchell's response clearly leaves the technological underpinnings of the thesis intact, this tends to represent e-topia as a thesis literally on the run, not burdened by the dead weight of the past, but light and agile enough to keep moving forwards.

The way the thesis proposes to achieve this is instructive and reveals a lot about its ultimate objective. This is because while e-topia is seen to mark a break with the past, dis-embedding insitu practices, 'churning' everything up and turning things around, all that follows in its wake is perceived as being integrated back into the increasingly carbon-based and silicate-permeated body of urban life that such developments pave the way for. That body of urban life which is now seen to provide cities with the platforms (computational frameworks, hardware, software, operational systems and programmes of coded languages), architects, planners, engineers and surveyors need in order for them to be smart in resizing communities and building the recombinant spaces also required to reconcile everything wrapped up in the displacement and relocation that this late modern process of globalisation produces.

What the emerging critique reveals

While the thesis may stress all the disembedding, churn, displacement and relocation wrapped up in the process of globalisation is virtuous, not least because it offers the possibility of resizing communities as recombinant spaces which are reconciliatory, unfortunately it does not have either the foundation nor superstructure to give this claim any more substance than the force of words used to advance it. For all the thesis has is the idea of putting an end to the trial separation of bits and atoms and call for the professional community of architects, planners and engineers to respond in a way never before thought possible. The problem with this lies with the tendency this well-narrated, deeply thought-out and heart warming, i.e. highly emotive and passionately argued, post-millennium thesis has to shout hard and loud about the need for such a response, but remain silent on the methodology required. That is, remain silent on either how the aforementioned professions can embed the intelligence needed for cities to be smart in resizing the communities required for their recombination to be reconciliatory.

These concerns are noticeable for the fact they tend to bring Mitchell's representation of e-topia as urban life, where it all comes together into question and casts doubt on the vision of the future the thesis sets out. What the emerging critique reveals is that if 'information society' is to fully grasp the opportunity e-topia offers to be reconciliatory, we first of all need to lose its environmental determinism; and secondly, see whether the techno-topian legacy that the thesis harbours is only loosely linked to the embedded intelligence of smart cities.

To decouple the determinism of the techno-topian legacy from the intelligence of smart cities and thereby retain the opportunity that e-topia has to be reconciliatory is a tall order. This is clear from Mitchell's own discussion on the city of bits surfacing in *High Technology in Low Income Communities* (Mitchell, 2001). For here Mitchell draws attention to the champions of this reconciliation – architects like Calthorpe (1993), Katz (1993) and Horan (2000) who are taking the lead and responding in a way never before thought possible. In a way, the likes of Calthorpe, Katz and Horan contend, that allows their particular brand of new urbanism to be reconciliatory in decoupling the determinism of the techno-topian legacy from the embedded intelligence of smart cities.

In many ways, though, it is evident what their particular brand of new urbanism offers is little more than a mirror image of what Mitchell's thesis does: that is side-step questions about the embedded intelligence of smart cities. The problem with this tactic lies with the methodological gap that it leaves for the professional bodies that the thesis is supposed to serve. For in its current state, the thesis leaves them in the unfortunate situation whereby they are unable to offer any critical insight into the embedded intelligence of smart cities, let alone take on the role of intelligent agents able to search out any opportunities there are for cities to be smart.

While problematic in itself, if the difficulties experienced were only methodological they may perhaps be manageable. The problem is they run deeper than this and their knock-on effects relate to more substantive issues surrounding the trajectory of Mitchell's thesis. In particular they relate to whether or not the response from the professional bodies in question is constructive in embedding the intelligence needed to resize communities and build the recombinant spaces required for cities to become smart. Furthermore, they raise questions as to whether all this resizing of communities and building of recombinant spaces turns on their use of ICTs and if the intelligence this generates is capable of reconciling one with the other.

This is an important point to stress because if the trajectory is not reconciliatory, then the question of whether or not all this intelligence can be seen to be smart, namely a progressive force for change, or merely a way of reproducing the status quo also surfaces. This matter arises because of what Graham and Marvin have referred to as the tendency towards 'splintering urbanism'. For according to their thesis, urban life is no longer able to support the sheer weight that material cities are expected to carry. This is because their scenario of 'splintering urbanism' has a vision of the future that is the direct opposite of the reconciliation that Mitchell sets out and which merely ends up being destructive, doing little more than building the capacity there is for the urban life of cities to be played out in even more dysfunctionally separated communities. Dysfunctionally separated communities whose multi-scalar resizing into recombinant spaces is simply not strong or resilient enough to keep everything stuck together.

In many ways this representation of splintering urbanism provides what can only be referred to as the antithesis of Mitchell's e-topia. An antithesis that goes a long way to search out, uncover and expose the other side of the thesis and do this in an attempt to throw light on the 'futureless' plight of the urban poor as low income communities living in the deprived quarters of the city. From this account of what is euphemistically referred to as the fragmentation of urban places as electronic spaces, it is evident that the problems with Mitchell's thesis on e-topia are as much substantive as methodological, the former holding the key to unlocking the latter.

The key thing to bear in mind with all of this is that everything, i.e. e-topia, or splintering urbanism, hangs on whether or not the embedded intelligence of cities is smart enough to meet the ecological integrity and equity that is required to stop communities literally falling apart and support their coming together. So, in this sense Mitchell is right, there is everything still to play for and the risk of the professions not taking action to resize communities as recombinant spaces is too great to contemplate. For if no action is taken, e-topia will remain little more than a speculative vision of urban life, culturally incapable of representing cities as leading examples of what is either intelligent or smart about such developments. At best only able to reproduce the status quo, or at worst contribute to the generation of even greater environmental degradation and social inequality.

The challenge

Accepting this, the question we currently face is not so much about whether to work with the vision e-topia sets out, or abandon it; but how best to meet the criticisms levelled at it. However, in accepting this, we subsequently come up against a challenge of some magnitude. This is because there is not only the question of how to meet the criticisms, but draw a line under them, move on and in effect do what others have as yet been unable to. That is do nothing less than show how the vision of e-topia is being developed as a means to demonstrate the potential that exists for the embedded intelligence of cities to be smart in allowing the resizing of low income communities to be constructive and build spaces which are ecologically sound and equitable.

The instruction we get from Mitchell on this matter appears in his statement on the materiality of the e-topia thesis in *Me++: The Cyborg-self and the Networked City*. For here Mitchell suggests it is not virtual versus physical, nor cyber versus meat space that is significant, but the 'networked intelligence being embedded everywhere' in cities that is the critical factor in them becoming lean, mean, green and smart. He suggests that this is because we are increasingly living out our lives in places 'where electronic information flows, mobile bodies and physical spaces interact in engaging ways', these 'occurrences' in turn pointing the way to the architecture of the twenty-first century.

Yet again, however, what this 'architecture of the twenty-first century' means for the urban life of cities is not clear and highly controversial. Especially if we contrast this post 2000 AD (literally, After Dematerialisation) account of late modern globalisation with the position taken by Mitchell (2005) in his more recent publication, *Placing Words*. For here all the significance previously lavished on the new subjectivity of the so-called cyborg-self and their civic existence as wireless bi-peds appears to be set aside. Here Mitchell suggests the node-based subjectivity of the cyborg-self is perhaps best understood not so much as the 'limb-like extension and sensory augmentation of urban life immersed in the programmable code of the city's de-privatised space', but as the context-specific language of social capital. The context-specific language of social capital supported by digitally inclusive regeneration platforms, whose embedded intelligence conveys what is smart about the informational content of low-level communications. Low-level communications whose embedded intelligence is smart because it conveys what is meant by the spoken words of everyday urban life and, for this reason, what all of this means in terms of the conversations, dialogues and discourses we routinely enter into with others as members of civil society.

Unlike Mitchell's declaration about the trial separation of atoms and bits being over, it appears that here in *Placing Words*, their materiality, i.e. tangible form and content, is something which is still very much in the process of being brought back together, so what their re-coupling means can be put on trial and understood in the everyday language of urban life as we have come to know it. That is to say, in the atomistic, bitty, fragmented, fragile, anxious and insecure encounters of everyday urban life and which we all share with other members of civil society.

Staking out the landscape

The landscape this chapter targets, aims to stake out and occupy, is the middle ground between the 'high-level' issues of the cyborg-self and those found at the 'grass roots' level of everyday linguistic practices (Mitchell's electronic space and Graham and Marvin's urban place, respectively). This is because it is here where 'what it all means' gets 'bottomed out' as part of the emerging discourse on the embedded intelligence of smart cities, represented as basic values (for example, ecological integrity and equity) underpinning the 'low-level' actions (planning, property development, design and construction activities) supporting their digitally inclusive regeneration (Curwell et al., 2004).

Approaching the question as to the viability of linking the embedded intelligence of smart cities in this way makes it possible to run vertically, digging deep into the planning and development of cities and using this as the basis to move horizontally. That is, as a basis to be constructive in using ICTs to build a platform of services which are capable of supporting a digitally inclusive regeneration (Cooper et al., 2005).

The outcome of this is a platform of services with the institutional depth needed for any such multi-scalar resizing to be constructive in building bridges between the spaces that have previously been divided. Those types of bridge-building exercises, this chapter argues, are of particular concern to the 'political body' of civil society and the public's call not for what has euphemistically become known as cyber, or meat-space, but a greater, more extensive and higher level of participation in decisions taken about the material basis of urban life. Greater in the sense that such participation is not just limited to urban land-use planning, but extends into the property development, design, construction, operation and use of buildings. More extensive in the sense the participation in question is not limited to the use of urban land, but 'scaled-up' and 'resized', using the step-wise logic of planning, property development, design and construction. That is, 'scaled-up' and 'resized' into a type of 'step-wise' logic which lifts us onto a stage no longer limited to matters of environmental concern, but instead able to reach out, extend beyond, consult with and include deliberations on matters about the social, environmental, cultural and economic future of urban life.

Taking the embedded intelligence of smart cities full circle

From this vantage point, it becomes possible to point towards what this all means. This in turn takes the investigation into the embedded intelligence of smart cities full circle and begins to answer the questions posed by the critics of Mitchell's thesis.

These questions will be answered by letting what follows outline urban life as we shall come to know it! That is to say, as a whole new landscape which seeks to sustain the urban life of cities through the planning, property development, design and construction of places, qualified in terms of the ecological integrity and equity of the villages and neighbourhoods making up the multi-scalar resizing of communities. Those multi-scalar and resized communities whose recombinant spaces are no longer alien, but now familiar enough – as villages and neighbourhoods – for the public to participate in constructing as cultural artefacts because the decisions taken about their design and layout have a bearing on the environmental and social future of their urban life.

The search for a representation of embedded intelligence which is the antithesis of an environmentally deterministic perspective of urban life and is smart in offering cities a role in 'making a place for low income communities' turns attention away from 'the electronic spaces of urban places' and towards the role of ICTs as (plat)forms of social capital (Halpern, 2005). For in turning attention towards the embedded intelligence of social capital, it subsequently becomes possible to recognise the critical role that the conversations, dialogues and discourses of everyday urban life play in the ability cities have to be smart in creating not only the norms, rules and expectations any multi-scalar resizing communities are subject to, but the recombinant spaces that are also part of this digitally inclusive regeneration.

This achieves what Mitchell makes clear is needed in his reference to Bretch's comments on the role of 'the radio as an apparatus of communication' (Mitchell, 2004). In particular, the need for the type of two-way information flow and multi-channelled communication that Mitchell believes to be the basis of the collaborative platforms, consensus building, ecological integrity, equity and democratic renewal required to rescale communities and 'size them up' as the recombinant spaces of urban life. This provides what Mitchell refers to as:

> a ... strategy that draws upon the lessons of the internet [which], is to think of [the platform] as a communal resource, like the old village commons, or the land available to the squatter community. [Because this means] anyone can use it as long as they follow a few rules ... (2005: 56)

The lesson learnt from this being to build such platforms from what Mitchell calls:

> the viral propagation of web links and email lists to support grass roots campaigning which are not constrained by distance. Blogs and online forum which substitute highly interactive discussion for the broadcasting of packaged messages ... (2005: 74)

While this points towards a strategy for the development of a collaborative platform, the lack of focus it offers on either the ICTs or the social capital of such structures means the statement about building consensus on the multi-scalar resizing of communities and the recombinant spaces of urban life is only concerned with the city's ecology and not the equity of resource distribution. Questions about equal access to and distribution of the opportunities such platforms offer to secure such services appear to be side-stepped. This is difficult because it means Mitchell's statement on the 'strategic response' tends to give out the wrong message. For while the message conveyed suggests it is relatively easy to construct such a platform and build the respective user services, experience teaches us this is anything but the case. Anything but the case for the simple reason the digital divide persists and the ICTs needed to bridge it are not available en masse for low-income communities to access as an intelligent way of being smart in using social capital to channel any such multi-scalar resizing.

It would appear that if we were to adopt this strategy the urban poor and dispossessed would remain sidelined and systematically excluded from what Mitchell refers to as the multi-scalar resizing of community and recombination of urban life which is currently being 'played-out' in the villages, neighbourhoods and districts of cities. This in turn producing not only a culture of no-go areas, but no-flow environments, without any economic future to boot!

The point made here is that in low-income communities the challenge is even greater because we are starting not just from such a low base line, but a level of urban development which is also excluded from the mainstream. Excluded from the mainstream for the reason the communities in question do not possess the embedded intelligence needed to be smart. That is: undergo such a multi-scalar resizing in line with the logic of the recombinant spaces which is being imposed upon them as a requirement of this digitally inclusive regeneration.

An instructive account of how a semantic web-based learning platform and knowledge management system capable of networking the types of electronically enhanced service developments outlined here can be used as a platform for digitally inclusive regeneration is reported on by Deakin and Allwinkle (2006, 2007) and set out in Deakin (2009a and 2009b).

This platform is drawn from an assessment of five leading city portals. This involved an evaluation of the services hosted on the city portals of Edinburgh, Dublin, Glasgow (Drumchapel), Helsinki (Arabianranta and Munala) and Reykjavic (Garoabaer). The evaluation of the portals included:

- a review of the learning and knowledge services these leading city portals offer stakeholders engaged in digitally inclusive regeneration;
- the benchmarking of their existing platforms against the users' knowledge transfer and capacity-building requirements;
- the selection of the ICTs best able to meet the semantic web requirements of this platform and develop as the natural language of a knowledge management system supported by a digital library;
- the integration of the aforesaid into a platform of e-government services available for members of the communities undergoing such a digitally inclusive regeneration to access as part of their multi-scalar resizing.

This work on digitally inclusive regeneration draws particular attention to the service developments underlying the so-called e-topia demonstrator and citizenship seen as supporting the active participation of communities in their resizing. In particular, in their resizing as the recombinant spaces of just such a process of democratic renewal (Deakin, 2009a and 2009b, 2010).

Conclusions

This chapter has carried out an extensive review of Mitchell's thesis on e-topia and found it wanting. It has pointed out that the problems encountered lie with the lack of substantive insight the thesis offers into the embedded intelligence of smart cities.

The chapter has identified that if the difficulties experienced were only methodological they would perhaps be manageable, but the problem is they run deeper than this and relate to more substantive issues which surround the trajectory of Mitchell's thesis. This is a critical insight of some significance because if the trajectory of the thesis is not in the direction of either the embedded intelligence of smart cities, or the ICTs and digitally inclusive regeneration platforms, then the question arises as to whether the whole notion of e-topia, the cyborg-self and their virtual communities, can be a progressive force for change, rather than merely a way for the embedded intelligence of smart cities to reproduce the status quo. Perhaps more importantly, this in turn begs the question as to whether attempts made to deploy the thesis will prove counter-productive. In that sense any attempt to govern the ICTs of digitally inclusive regeneration platforms shall fail because they are unable to be reconciliatory in bridging social divisions and for this very reason do little more than merely add to the inequality of the ecological degradation that is already being experienced.

This unfortunate scenario is drawn from what Graham and Marvin have referred to not as e-topia but splintering urbanism, because, according to their thesis, the citizenship underlying these communities is no longer able to carry the sheer weight of the material that such a cybernetic-based networking of intelligence is supposed to support. This is important because their scenario has a vision of the future that is the direct opposite of what Mitchell represents and a knowledge base which ends up with cities being not so much smart as a dumping ground for social inequalities which are themselves ecologically destructive. That is, which ends up doing little more than building the capacity that exists for storylines on the mythology of community governed by a multi-scalar resizing not strong enough, or sufficiently resilient, for the recombinant spaces this produces to stick everything back together.

Their representation of splintering urbanism provides what can only be referred to as the antithesis to Mitchell's e-topia. An antithesis that is important because it goes to extreme lengths in searching out, uncovering and exposing the other side of cybernetic-based intelligence which currently lies hidden, and does so by throwing light on the plight of low-income communities living as the urban poor in deprived quarters of the city. From this it is evident that the problems with Mitchell's thesis on e-topia are as much sub-stantive as methodological, the former holding the key to the latter.

In substantive terms this chapter has gone very much against the grain, arguing that our current understanding of embedded intelligence, smart cities and the ICTs of digitally inclusive regeneration puts us on the verge of cul-tivating a new environmental determinism. An environmental determinism which this time around is cybernetic, based on the embedded intelligence of knowledge-based agents underpinning the networking of smart cities and the digitally inclusive regeneration platforms they support. To avoid repeating this mistake (yet) again, attention has been drawn to the work of Graham and Marvin and the spaces that their radical democratic, i.e. egalitarian and eco-logically integral, account of the situation opens up for a much more eman-cipatory view of the intelligence embedded in those knowledge-based agents smart enough to meet these requirements.

Those knowledge-based agents smart enough to meet these requirements and do so by way of and through the social capital underlying and giving rise to the norms, rules and values of such developments. In particular, the social capital that underlies the embedded intelligence of smart cities and which their knowledge-based agents (architects, planners, engineers and surveyors) in turn support by hosting them as services found on digitally inclusive regen-eration platforms. Digitally inclusive regeneration platforms whose equity, ecological integrity and democratic renewal govern over the modernisation of villages and neighbourhoods their step-wise logic pave the way for.

In ignoring these warnings and being unable to learn the lessons such a critical reworking of the thesis offers, the strategy Mitchell adopts must be seen as suspect. Not only because the vision and scenarios it advances also end up side-stepping such concerns, but for the reason it replaces the agonies of equality and ecological integrity with the 'gnostics' of 'new age' wordings, centred around storylines about the quality of life. Storylines that spell out, write about and communicate the experiences of those organisations that are wrapped up in such developments.

The strategy advocated for adoption by this chapter is not grounded in such rhetoric. Its vision of e-topia builds instead on the messages the likes of Graham and Marvin advance by turning the tables and agreeing that while words offer the possibility of 'bringing what it all means back together', actu-ally turning things around lies not so much in the words, as it rests with the semantics of the syntax and vocabulary governing the regenerative storylines that emerge from the citizenship (cyborg-civics) of virtual communities and the degree to which they manage to overcome the divided antagonisms of the excluded.

This way it becomes possible for the multiplied memory and infinite mind of the cyborg civic and their tribe-like culture, not to bemoan the nomadicity of wireless bi-peds, but actively celebrate the creativity of the virtual com-munities emerging from the digital-inclusive nature of such regenerative storylines. In particular, celebrate their capacities to be both analytical and synthetic and the opportunity this in turn creates for virtual communities to use the collective memory, wikis and blogs of electronically enhanced services as a means to bridge social divisions.

References

Brecht, B. (1932) The Radio as an Apparatus of Communication, in John G. Hanhardt, ed., *Video Culture: A Critical Investigation*, Rochester, NY, Visual Studies Workshop Press.

Calthorpe, P. (1993) *The Next American Metropolis: Ecology, Community, and the American Dream*, New York, Princeton Architectural Press.

Cooper, I., Hamilton, A. and Bentivegna, V. (2005) Networked Communities, Virtual Organisations and the Production of Knowledge, in Curwell, S., Deakin, M. and Symes, M. eds, *Sustainable Urban Development, Volume 1: The Framework and Protocols for Environmental Assessment*, Oxon, Routledge.

Curwell, S., Deakin, M., Cooper, I., Paskaleva-Shapira, K., Ravetz, J. and Babicki, D. (2004) Citizens' Expectations of Information Cities: Implications for Urban Planning and Design, *Building Research and Information*, 22: 1.

Deakin, M. (2009a) The IntelCities Community of Practice: The eGov Services Model for Socially-Inclusive and Participatory Urban Regeneration Programs, in Reddick, C., ed., *Handbook of Research on e-Government*, Hershey, IGI Publishing.

Deakin, M. (2009b) A Community-based Approach to Sustainable Urban Regeneration, *Journal of Urban Technology*, 16 (1): 191–212.

Deakin, M (2010) Review of City Portals: The Transformation of Service Provision under the Democratization of the Fourth Phase, in Reddick, C., ed., *Politics, Democracy and e-Government: Participation and Service Delivery*, Hershey, IGI Publishing.

Deakin, M. and Allwinkle, S. (2006) The IntelCities Community of Practice, *International Journal of Knowledge, Culture and Change Management*, 6 (2): 155–162.

Deakin, M. and Allwinkle, S. (2007) Urban Regeneration: The Role Networks, Innovation and Creativity Play in Building Successful Partnerships, *Journal of Urban Technology*, 14 (1): 77–91.

Graham, S. and Marvin, S. (1996) *Telecommunications and the City*, London, Routledge.

Graham, S. and Marvin, S. (2001) *Splintering Urbanism*, London, Routledge.

Halpern, D. (2005) *Social Capital*, Bristol, Polity Press.

Horan, T. (2000) *Digital Places: Building Our City of Bits*, Washington, DC, Urban Land Institute.

Katz, P., ed. (1993) *New Urbanism: Towards an American Architecture of Community*, New York, NY, McGraw-Hill.

Mitchell, W. (1995) *City of Bits: Space, Place, and the Infobahn*, Cambridge, MA, MIT Press.

Mitchell, W. (1999) *E-topia: Urban Life, Jim – But Not as We Know It*, Cambridge, MA, MIT Press.

Mitchell, W. (2001) Equitable Access to an On-Line World, in D. Schon, B. Sanyal and W.J. Mitchell, eds, *High Technology and Low-Income Communities*, Cambridge, MA, MIT Press.

Mitchell, W. (2004) *Me ++: The Cyborg-self and the Networked City*, Cambridge, MA, MIT Press..

Mitchell, W. (2005) *Placing Words: Symbols, Space, and the City*, Cambridge, MA, MIT Press.

Williamson, Imbroscio, D. and Alperovitz, G. (2003) *Making a Place for Community*, New York, Routledge.

Part II
Modelling the transition

7 Smart cities

A nexus for open innovation?

Krassimira Paskaleva

Introduction

This chapter critically reviews current European trends on smart cities in the context of open innovation. It draws from analyses of key European Union (EU) programmes, four international projects and related activities. These initiatives are framed by the EU's strategic policies on competitiveness and innovation, smart cities, the Future Internet and living labs (LLs), with the aim of fostering smarter, sustainable and inclusive cities. The analysis shows that a new approach to open innovation is emerging which links technologies with people, urban territory and cities, and which is likely to be increasingly influential. It is suggested that using open innovation for sharing visions, knowledge, skills, experience and strategies for the planning and design of the delivery of services, goods and policies is effective, efficient and sustainable. However, consistent frameworks, principles and strategic agendas are necessary to bind these elements together.

The smart cities agenda

Amid profound economic, social and technological changes, caused by globalisation and integration processes, cities in Europe and around the world are faced by the challenge of reconciling competitiveness with long-term sustainable development. In the era of the digital economy, the performance of cities is influenced not only by their physical infrastructure, but more and more so by their knowledge and social capacity ('intellectual and social capital'). This later form of capital has recently been considered most critical for achieving sustainable and competitive cities. Against this background, the construct of the 'smart city' has emerged as a strategic agenda over the past ten years, emphasising the increasing importance of information and communication technologies (ICTs) for encompassing modern urban development factors in a common framework and for profiling cities' competitiveness based on their social and environmental capital (Caragliu et al., 2009). The significance of these two factors in fact is known to distinguish 'smart cities' from their more technology-burdened peers known as 'digital' or 'intelligent' cities.

Over the course of the last several years and in the context of building the digital economy, the 'smart cities' agenda has gained a real momentum in the countries from the European Union (Komninos, 2002; Paskaleva, 2009; Deakin, 2010). Other international organisations, such as the OECD, have also added to the agenda, by tasking ICTs with the job of realising strategic urban development goals, such as sustainability and improving the quality of life of the citizens (OECD – EUROSTAT, 2005). Similarly, many cities have used the 'smart city' term to profile themselves as prosperous and well endowed, in a variety of circumstances: For instance, the Amsterdam Smart City initiative emphasises the importance of collaboration between the citizens, government and businesses, so as to develop smart projects which will 'change the world' by saving energy; Southampton City Council uses smart cards to stress the importance of integrated e-services; the City of Edinburgh Council has formed a smart city vision around an action plan for government transformation; and the Malta Smart City promotes a business park to promote sustainable economic growth. IBM, Siemens and ORACLE have also formed their visions of smart cities. Along with this, a number of EU research and policy projects have also emerged to deal with the challenges that smart cities pose.

A recently concluded pan-European research project, IntelCities (2009), also concluded that governance, as a process and outcome of joint decision making and action, has a leading role to play in building the smart city and that cities should develop collaborative digital environments to boost local competitiveness and prosperity by using knowledge networks and partnerships, integrated e-services and governance (Paskaleva, 2009; Deakin, 2009). At present a Smart Cities INTERREG project (SmartCitiesProject, 2011) is also developing an innovation network to deliver better e-services in the North Sea region (Deakin, 2010; Deakin et al., 2011). All this follows a 2009 Smart Cities Future Conference in Manchester (UK), which endorsed open innovation and effective collaboration between stakeholders as the critical factors for starting the journey towards the future smart city.

Yet, as Hollands (2008) has recently revealed, cities all too often claim to be smart, without defining what this means, or offering the evidence to support such claims. Moreover, smart-er cities appear to be simply 'wired cities', whereas: 'progressive(ly) smart[er] cities must seriously start with people and the human capital side of the equation, rather than blindly believing that IT itself can automatically transform and improve cities' (Hollands, 2008: 315). And as he goes on to conclude: 'the critical factor in any successful community, enterprise, or venture, is its people and how they interact' (ibid.: 306). It is around this viewpoint that smart cities can be progressive because they use digital technologies not to hardwire themselves but to be socially inclusive, foster good governance and create better services which improve the quality of life of the citizens, with an outlook to long-term sustainability and competitiveness.

Taking this vision of smart cities forward, the chapter reflects upon the current trends and understanding of what it means for urban administrations, policy makers and businesses in European cities to be smart, both now and in the future. The role of the smart city as a nexus for open innovation is marked out for special attention, as this theme and strategy have become focal points in current discussions amongst member states in debates about the Future Internet, living labs and innovation and competitiveness-driven (urban) development. By conducting a critical review of some high-profile programmes and initiatives on smart cities, the emerging trends are explored and insights are drawn about the challenges that 'smart-er cities' pose. The analysis is based on four smart cities projects and the relevant EU programmes. They have been chosen because collectively they provide a strategic overview of what Europe expects of smart cities. The analysis also responds to the 'quest of the research and academic communities', as Holland (2008) puts it, to identify the defining components, critical insights and institutional means by which to get beyond the self-congratulatory tone of smart cities.

The chapter is structured around five main sections. As this introduction has set out the rational for the study, section 2 discusses the relevant research background. Section 3 presents some major European projects on smart cities. Section 4 follows up with a discussion about the emerging trends and draws insights about the driving stimulus behind building smarter future cities in Europe. The conclusion sums up the outcomes of the study against its objectives and identifies strategic questions in building a smart city whose system of open innovation is sustainable, competitive and inclusive.

Theoretical background

In the course of the last decade, the concept of the smart city, considered by many as the new century's stage of urban development, has become trendy in the policy and business arenas (Komninos, 2002). Academic interest in the subject has also grown. Here mainstream approaches tend to deal with issues relevant to the use of ICT in making cities more technologically advanced. Accordingly, the availability and quality of the ICT infrastructure have become the determining factors for many cities seeking to brand themselves 'smart'.

In contrast to this predominantly technical approach to smart cities, a growing number of studies have recently emerged which suggest that it is in fact environmental, cultural and social capital which now drives urban progress. Factors like the capacity of the human (Berry and Glaeser, 2005; Glaeser and Berry, 2006), relational capital (Paskaleva, 2009) and the role of higher education, skills, creativity and talent (Shapiro, 2006; Winters, 2010; Mellander and Florida, 2009) have all emerged as the main drivers of urban development. Within these studies the local quality of life factor also becomes more and more important in determining population clustering, amongst other modern urban phenomena, and the overall significance of territorial amenities is emerging as one of the major components of urban attractiveness and development (Rappaport, 2009).

But despite the significant advance in ICT and urban research moving beyond the technology of urban growth, the construct of the smart city still remains ambiguous. Recent research carried out for the development of the Wikipedia website on smart cities has revealed that up until now smart cities have been generally identified along six main dimensions: smart economy; smart mobility; smart environment; smart people; smart living; and smart governance (Wikipedia, 2011). It has also been revealed that the main characteristics of a smart city can be grouped in three main categories:

(a) *Exploitation of networked infrastructures:* to improve efficiency and enable the social, environmental and cultural attributes of urban development (Hollands, 2008; Nijkamp, 2008), with the term infrastructure signifying business, leisure and lifestyle services, housing and ICTs (satellite TVs, mobile and fixed phones, computer networks, e-commerce, internet services), and a smart city meaning a wired city as the main development model and connectivity as the source of growth (Komninos, 2002).

(b) *A vision and strategy for creating the competitive city*, with the smart city taking the opportunities ICTs offer to increase local prosperity and competitiveness, approaches here varying from stressing the importance of the multi-actor, multi-sector and multi-level urban perspective towards competitiveness and sustainability (Paskaleva, 2009; Odendal, 2003), to signifying the presence of a creative class, dedicated to a high-quality urban environment, educational achievement and multi-channelled access to ICTs serving either the public administration of cities (Caragliu et al., 2009) or business-led urban development (Hollands, 2008).

(c) *An approach to sustainable and inclusive cities*, placing the main weight on the social capital of urban development. Here smart cities know how to learn, adapt and innovate and the focus is either on inclusive public services (e.g. Southampton's smart card) or involving citizens in the co-design of services (Southampton City Council, 2006; Deakin, 2007; Deakin and Allwinkle, 2007; Lombardi et al., 2009). Sustainability, in this sense, is considered as the very strategic element of the smart cities agenda, and achieving social, cultural and environmental sustainability through participation of the public in local decision making is seen as key to the democratisation of such governance (Caragliu et al., 2009; Deakin, 2010).

But despite the growing attention, definitions or systematic models of the smart city remain scarce and the debate on its principles and strategic policy agenda remains unresolved. At the same time, cities are continuing to grow, urban infrastructure and resources are becoming scarcer, and the state of the urban environment is determining most people's quality of life, which suggests that urban evolution should be driven by clear agendas about the future of sustainable cities. Understanding the factors that make cities and urban livelihoods smart, yet sustainable with a better quality of life, appears decisive for future debates on the smart city.

Although the smart-er city is generally considered an outcome of smart cities practices, this chapter seeks to broaden its scope by looking at the smart city as an 'activator' of change through exploring relevant open innovation processes. With the latter emerging as a powerful driver of this new paradigm shift, no general or explicit attention has yet been given to a number of important questions:

(a) How does open innovation shape the way cities become smart?
(b) To what extent is the way in which this operates 'transformative'?
(c) Can the dynamics of open innovation systems generate new communities, organisations and processes?
(d) Can such innovation systems help cities cope with the challenges of implementing new technologies everywhere and for everything?

While some of these questions have already been raised and many of them have now found their way into EU strategies, as yet few answers have been forthcoming.

The EU smart cities agenda

The smart cities theme has emerged in several European programmes, but its importance has recently increased in the landscape of the Future Internet, the main argument being that smart cities can serve as a catalyst for Future Internet research, as they form dense social ecosystems which heavily rely on internet technology and which in turn heavily influence social interactions. It has also been argued that it is in smart cities where the initial impact of the Future Internet will be most visible to European citizens, and where direct feedback from EU citizens on Future Internet technology and applications can be obtained (Lemke and Luotonen, 2009). As a result of this policy, a number of new initiatives have emerged in the last couple of years, such as the FIREBALL and FIRESTATION co-ordination and support actions, the Smart Santander smart city experimental facility and several smart city pilots funded under the ICT Policy Support Programme (PSP) of the European Commission.

A number of EU programmes have also highlighted the importance of the smart city for Europe's Digital Agenda. The i2010 (EC, 2010a) initiative emphasised three main pillars of the digital society: (i) developing the Single European Information Space, in order to promote an open and competitive internal market for information society and media; (ii) further strengthening of innovation and investment in ICT research, so as to promote growth and more and better jobs; and (iii) ensuring an Inclusive European Information Society, which is consistent with sustainable development and is able to prioritise better public services and quality of life.

Following on from this statement, the Directorate General Information Society's Competitiveness and Innovation Programme has launched the ICT

PSP Programme for stimulating innovation and competitiveness through the wider uptake and enhanced use of ICTs by citizens, business (particularly SMEs) and government, in which user-driven innovation is linked to internet-enabled services in smart cities.

This 'innovation strategy' was also at the heart of i2010 and continues to drive policy activities. The importance of the role played by the end users in the digital society is particularly recognised. But it is clear that as services and applications develop, they will need to be sustained through user-driven open innovation in order to be scalable and replicable at a mass-market level. It is also recognised that as user-generated content and co-created applications grow, they will need to be linked to new and innovative business models which will ensure effective implementation and sustainability. Therefore bringing together Future Internet technologies with living labs methodologies and practices has been chosen as a viable way to advance the development of e-services (EC, 2010f). However, despite all the efforts during the last 15 years to improve government services, transactions and interactions with European citizens and businesses, technical and procedural limitations have combined to prevent cities from truly harnessing the full power of ICT to collaborate, create and deliver genuinely 'smarter' citizen- and business-centred services. The current economic crisis combined with growing citizen expectations is placing increased pressure on cities to overcome these barriers and provide better and more efficient service infrastructures. Innovative ICT solutions – particularly those created in the user-driven, open innovation environments of living labs – appear to be key in helping European cities overcome such barriers.

Therefore the new forward-looking 2020 Strategy emphasises three main types of growth – *smart growth* (fostering knowledge, innovation, education and digital society), *sustainable growth* (making our production more resource efficient while boosting our competitiveness) and *inclusive growth* (raising participation in the labour market, the acquisition of skills and the fight against poverty). Current opinion on which of the three objectives poses the greatest challenge suggests that smart growth and *innovation* is the most pressing matter within Europe's 2020 Strategy (EC, 2010b). The latter has been taken up by the latest EU commitment to support the development of living labs as an environment for open innovation. The European Network of Living Labs demonstrates the desire of the living labs cities to share their experience in Europe and globally. National and regional networking of living labs is also increasingly taking place, e.g. in UK, Belgium and Finland. The main idea of the living lab concept, on which this paper builds, is keeping the users continuously involved in making better products and services while their expectations are continuously monitored and reflected upon in a systematic process.

Other EU programmes are also influencing the smart city agenda. On a spatial scale, the Lisbon and Gothenburg strategies are particularly relevant and take advantage of emerging Future Internet technologies to achieve competitiveness and sustainability in a complementary symbiosis, rather than as a trade-off. The Territorial Agenda, URBACT Programme and the Leipzig Urban Charter both focus attention on cities and emphasise the need to build on their potential as centres of knowledge and as sources of growth and innovation. On a sectoral level, the Future Internet initiative aims to redesign the internet, taking a broad multidisciplinary approach, to meet Europe's societal and commercial ambitions. Together these programmes are shaping the future of the smart city agenda.

The role of open innovation in the smart city

Although the ideas and discussions date back to the early 1960s, it was Henry Chesbrough who first promoted the idea of 'open innovation' in 2003 as a paradigm which assumes that when advancing their technologies, firms can and ought to use external ideas as well as internal ideas as pathways to the market. Most recent debates, however, suggest that 'open innovation' should not just refer to industry but also to the ways government and other institutions work and collaborate with society. This emerging notion of open innovation, based on networking and inter-institutional relations, appears highly relevant to the paradigm of the 'smart city' that is anticipated in the present study.

As the critique of the predominantly technology-driven approach to smart cities is increasing, there is a strong need to align innovation policies with the goals of urban development. Smart cities also require 'smart citizens' if they are to be truly inclusive, innovative and sustainable. The promise of the information society to create new ways of empowering people to play a fuller and more equal role in emerging governance systems through their access to dynamic internet-enabled services is also proving to be challenging, because not everyone is getting equal access to the skills and opportunities needed. Previous EU initiatives, particularly those focusing on e-government and e-inclusion, have also done much to tackled the 'digital divide', but only to find that the persistent inequalities blighting many urban neighbourhoods undermine citizen participation. This calls for a new approach in which the focus is first and foremost on citizen empowerment as an essential catalyst in creating a new paradigm to transform the dynamics of data flows, information management and service development towards the smart city. The potential of new bottom-up approaches based on user-generated content, social media and Web 2.0 applications opens up vast possibilities for a new interpretation and understanding of spatial differences and local effects, seen through the experiences of the citizens themselves, these in turn leading to new forms of citizen empowerment. The latter offers the potential for citizens to build not just the social capital, but the capacities required to become co-creators and co-producers of new and innovative services with the means to ensure that they are delivered in more effective and inclusive ways, taking full advantage of new internet-based technologies and applications (Cahn, 2001).

Developing collaborative processes between local 'smart citizens', government and developer communities will evidently support and enhance the process, which brings up the idea of 'co-production' of goods and services as core to 'open innovation' for the 'smart-er city'. Amidst the most recent crisis of reform of public services in the United Kingdom, a ground-breaking NESTA report (Boyle and Harris, 2009) went on to reveal that 'co-production' offers a new way for citizens to share not just in the design but also in the delivery of services and contribute their own wisdom and experience in ways that can broaden and strengthen services and make them more effective. As a result, co-production is developing as a practical agenda for system change in the UK, based on four key principles:

*   recognising people as assets
*   valuing work differently
*   promoting reciprocity
*   building social networks.

Forming partnerships between professionals and the public yet again appears crucial for improving the effectiveness and efficiency of local services. Establishing systematic and long-term collaboration between front-line practitioners and developers can help create a more positive environment for the 'co-production' of local services, but in policymaking as well, so as to make the system more responsive to community needs. Internet-based technologies and e-services provide endless opportunities for stimulating 'co-production'. In return, the latter can provide new opportunities for securing citizens' engagement and active involvement in the process of developing 'smart services' which can help to accelerate the uptake of these technologies and services. This virtuous circle is then capable of enhancing cities' ability to grow and sustain 'innovation ecosystems' and, through this, develop more inclusive, higher quality and efficient services. The added value for the users is that they have a real incentive to become more involved as co-producers, as well as users, of the content and services available in the emerging smart city through having access to new skills, employment possibilities and better quality of life. It is these possibilities which can then make such approaches more sustainable, by embedding the proactive involvement of citizens in all aspects of designing and delivering services, thereby giving a new rationale both to those citizens and to the public authorities responsible for providing these services. That rationale is becoming known as PPPP – public-private-people partnership – an approach which is viable and desirable in the smart city (SMARTiP DoW, 2010).

Using the living labs approach for smart city innovation

Innovation policies that support and foster innovation processes strategically are perceived as being crucial for increasing urban competitive advantages in the future. For ICT innovations, in particular, more open and networked forms of collaboration between industrial, governmental, academic and user stakeholders in the innovation process have been identified as a serious policy challenge. Yet, experience has shown that such open or networked innovation should not be interpreted in terms of a naive or ideologically driven concept, but rather in terms of a concrete solution for dealing with complex and systemic innovation of ICT products and services which are composed of many complementary components, as well as for dealing with the fundamental unpredictability of ICT usage.

Against this background, the living lab (LL) approach has grown rapidly in Europe as an outgrowth of William J. Mitchell's original concept at the beginning of this century and as a way of involving city dwellers more actively in urban planning and city design (Mitchell, 2005). Here the living lab is seen as a platform for implementing an open innovation model to pilot different initiatives towards the Europe 2020 perspective of well-being and sustainability. The LLs are commonly defined as user-driven innovation ecosystems, based on a business-citizens-government partnership which enables users to take active part in the research, development and innovation process (EC, 2010d). Partners include cities, municipalities, innovation agencies, universities, large industrial partners, SMEs, citizens and so on. Benchmarks of living labs are environments in which technology is given shape in real life contexts and where (end) users are considered 'co-producers' (Ballon et al., 2007; Jensen, 2007). An ecosystem is established in which new products and services are created, prototyped and used in real-time environments. Thus, users are not treated as objects in the innovation process or as mere customers, but as early stage contributors and innovators (Wise and Høgenhaven, 2008).

Since 2006, living labs have rapidly grown throughout Europe. Networking activities have also been established to share principles and best practices and some national and European projects (including Laboranova, Ecospace and C@R) have explored issues dealing with creating awareness, developing tools and methods, as well as learning from best practices. The European Network of Living labs (ENOLL, 2010) has been at the centre of this new movement among cities, which in 2010 had 250 living labs across Europe on both the interregional level (e.g. Nordic-Baltic Network of Living Labs), as well on the national level (Italian Network of Living Labs, Finnish Network of Living Labs and UK Network of Living Labs). Recent efforts have also shifted towards using the results of this cross-border collaboration to connect smart cities, but the question of how open innovation can become a true and effective instrument for making cities smarter remains a challenge. Core to finding the adequate solutions, as the previous discussion has shown, there is the matter of both designing and adopting a smart city model on which cities can become smarter, sustainable and inclusive in a systematic and cohesive way. The following analysis attempts to shed light on the approach by exploring the open innovation adopted by living labs as an ecosystem for smart cities.

Smart city trends in Europe

This section focuses on the main approaches and objectives of four recently launched policy support projects in Europe. The analysis aims to show how ideas and strategies are being strategically shaped across the areas of smart cities, the Future Internet and living labs.

The SMARTiP project: Smart Metropolitan Areas Realised through Innovation and People

SMARTiP builds on the philosophy that developing 'smart citizens' within a network of 'smart cities' can be an important catalyst for 'smart growth', one that will curb the inequalities in smart citizens and public services. Headed by the Manchester Digital Development Agency, this 13-partner initiative takes a holistic approach to e-government, so as to tackle various interconnected policy agendas simultaneously, and addresses the need to recognise the role of people in achieving this in a sustainable and fairer way. Here it is assumed that socio-economic disadvantage goes hand in hand with the 'digital divide', and access to the internet, or what is sometimes termed 'networks of opportunity' (Graham and Marvin, 1996), is a basic requirement for a full realisation of citizens' participation in a democratic society (Mossberger et al., 2008). Thus the incentive to widen access to technology, particularly in areas of deprivation, seems highly justified. For without sufficient demand for online services from all strata of society, the cost-benefits rooted in ICT-driven automation shall remain unlocked (SMARTiP, 2010).

Despite the progress in e-government, it has been widely acknowledged that the promise of the 'democratising' and 'empowering' impacts of networked ICTs has not been fully realised and there still remains a fundamental gap between the vision and delivery of government e-services. Among other things, the casting of the 'citizen' as a 'customer' has hampered more creative developments in government–citizen relationships. And whilst the public sector is yet to fully exploit the opportunities offered by online technologies, Web 2.0 has taken internet users, private individuals as well as businesses, by storm. What is key to 'all things 2.0' is that individuals drive the content development; they broadcast, share and consume content of their own choosing and often creation. For government, there is no clear or coherent strategy yet for the use of Web 2.0. Instead, a raft of initiatives has emerged incrementally, seeking to 'crowd-source' public policy, or to connect with citizens via online platforms. Recent examples of this are two centrally run campaigns in the UK: 'Show us a better way' – to hear your ideas for new products that could improve the way public information is communicated (UK Cabinet Office, 2008) – and 'Building Democracy', which supports projects that develop new ways of helping people participate in public discussions and influence government policy (UK Ministry of Justice, 2008).

Going beyond notions of value for money, the social inclusion agenda referred to in the SMARTiP project opens up an important dimension to the technology that should be harnessed by local government. The argument made is that democratic participatory values should be incorporated into service technologies adopted by local government. Whilst Web 2.0 is diverse and can encompass all forms of participation, facilitating smarter design of public services, as well as offering recognition of residents' own concerns, remains key to the challenges ahead. It is about building a local online presence, an enhanced sense of place and improving services in ways which help to reduce spatial inequalities that will drive the smart city agenda in the future.

Therefore creating a digital community with smart citizens and smart developers is the main instrument for achieving the goals of creating new, dynamic relationships in the smart city. The common interests can be drawn from any area of the digital and non-digital world. SMARTiP aims to build the 'bridges' needed to connect 'digital communities' together, enabling developers and citizens to find dynamic and imaginative ways to interact and to build wider 'communities of interest' by drawing inspiration and experience from the open source community and the social economy world. The project is based on the experience gained from digital communities in Manchester with the Manchester Living Lab, building upon the work being undertaken to create cross-border collaboration, through specific projects such as APOLLON and more generally through ENoLL. The new process of co-production is employed across the three main themes, (i) smart engagement; (ii) smart environments; and (iii) smart mobility (Figure 7.1).

As a result, SMARTiP will advance developments in several cross-cutting policy lines – smart cities which need smart citizens, innovation driven by the users, service co-creation and co-production, user involvement as producers, empowering citizens to become part of the innovation process, adequate social capital for the Future Internet, and new innovative business models to support implementation. This means new collaborations have to be built, so cities can understand innovation, innovators can understand cities and citizens can in turn become really engaged users, who are not merely content providers, but both the producers and deliverers of services. It is anticipated that this mechanism shall build a platform for long-term collaboration between smart cities, raising the profile of living labs and Future Internet methodologies, creating open 'citizen developer' communities, establishing PPPPs, co-production in the design and delivery of services, and the transferable application of services with adequate business models (Carter, 2010). The project involves a cross-disciplinary team of five city regions – Manchester, Ghent, Oulu, Bologna/Emilia Romagna, and Cologne – technology developers, living labs, the University of Manchester and a locally based NGO known as People's Voice Media.

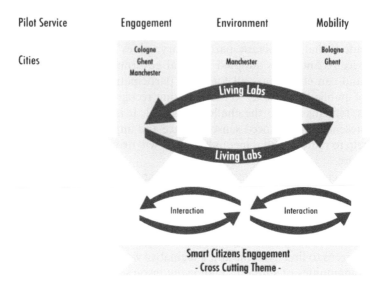

Figure 7.1 SMARTiP pilots' focus and methodology

Perifèria: networked smart peripheral cities for sustainable lifestyles

Perifèria's main goal is to deploy convergent Future Internet (FI) platforms and services for the promotion of sustainable lifestyles in and across emergent networks of smart peripheral cities in Europe, with a specific capacity for green creativity. Its Open Service Convergence Platform, an 'Internet by and for the People', assumes a social information architecture, integrating sensor networks, real time 3D and a set of mobile location-based services with the Future Internet paradigms of Internet of Things, Internet of Services and Internet of People. Specific technologies – the results of previous research initiatives, emerging devices and platforms, and ad hoc mash-ups, designed and implemented on the spot – come together to converge in a common living-lab-like setting for the co-creation of next-generation personal and collective services (Perifèria, 2010).

At the heart of Perifèria's convergent platform model is social interaction, which occurs at the 'run-time' moments in which the infrastructures and services are jointly and dynamically discovered, invoked and composed. Here user-generated content is considered the main driver of social interaction, with the latter occurring in different urban settings and conditions. Taking this notion forward, the five project pilots – (i) user-generated media for inter-cultural dialogue and civic interaction (Malmo, Sweden); (ii) traffic and transportation-related information (Bremen, Germany); (iii) strategic planning (Athens, Greece); (iv) cultural and natural heritage (Genoa, Italy); and (v) e-government services to citizens and businesses (Palmela,

Figure 7.2 Perifèria's Future Internet concept

Portugal) – are designed to analyse developments across diverse cultural, institutional and territorial frameworks. The key belief is that convergence of Future Internet platforms occurs through social interaction in concrete situations, in an 'Internet by and for the People' which is a discovery-driven rather than functionalities-driven centripetal aggregation of the main Future Internet paradigms. These include:

- The *Internet of Things (IoT)* – the ambient mobile-enabled infrastructure of sensors and connectivity which is fully networked and context-aware.
- The *Internet of Services (IoS)* – the service-oriented environments that finally break away from a functional logics approach to embrace a participatory, smart gaming and simulation approach to service discovery and provision.
- The *Internet of People (IoP)* – the emergent social architectures of relations, transactions and learning, using semantics of time and place as well as semantics of inter-personal situations. (See Figure. 7.2.)

The living lab approach is assumed to shift technology R&D out of the laboratory and into the real world in a systemic blend of technological and social innovation. This occurs through a 're-negotiation' of specific city infrastructures (named 'Urblets') and patterns of behaviour (named 'Behavlets') driven by Future Internet possibilities through Serious Games.

Within this setting, five archetypical 'arenas' – specific urban settings or innovation playgrounds, with defined social features and infrastructure requirements – emerge. These are the spaces where co-design and service integration processes unfold:

- *Smart Neighbourhood*: where media-based social interaction occurs
- *Smart Street*: where new transportation behaviours develop

- *Smart Square*: where civic decisions are taken
- *Smart Museum and Park*: where natural and cultural heritage feed learning
- *Smart City Hall*: where mobile e-government services are delivered.

A 12-partner team, headed by Alfamicro Sistema de Computadores Lda (Portugal), is carrying out the project, putting together industry and research experts, together with five city administrations from Europe.

EPIC: delivering effective smart city services across Europe

EPIC combines innovation ecosystem processes, e-government service applications and new cloud computing technologies to create a scalable and flexible pan-European platform – The European Platform for Intelligent Cities – for innovative, user-driven public service delivery through user-driven open innovation, connected smart cities and web-based advanced services. The EPIC platform combines latest technologies, from a semantic engine with 3D geo-locating capacities linked to the Future Internet, and offers the possibility for existing city innovation ecosystems to enhance their R&D process and enable them to deliver smarter city services.

This will enable (i) local SMEs to rapidly prototype scalable new user-driven solutions, (ii) innovative public administrations to test and deploy them, and ultimately (iii) cities across Europe to access and use them. By providing access to a market-leading shared infrastructure that facilitates rapid prototyping and testing, EPIC aims to drive innovation forward. The EPIC service platform combines city applications leveraging living labs and smart cities service delivery innovations, such as Relocation Service, Urban Planning Service and Smart Environment Service, to implement the new intelligent cities platform.

The idea is to reach out to cities, living labs, businesses and other stakeholders, to collaborate in accelerating innovation and smart service delivery. The hypothesis is that cities will be able to become smarter if they utilise the EPIC platform. A roadmap will guide their efforts for improving service delivery to achieve the benefits of smart working. Living labs methodology is used for testing and validating Apple iPhone services thorough engaging the citizens, and SMEs and cities plug existing and new co-designed web-based services into the open EPIC platform, so that aspiring cities, such as Tirgu Mures in Romania, can easily connect to the platform and use them. The added value to cities and living labs will be a new smart service delivery infrastructure in a scalable and cost-efficient manner, with easy access to innovative applications from across Europe. It will meet the increasing desire of citizens and business to access localised, user-centred government services jointly with other private sector offerings, i.e. quicker, faster and more personalised access. 15 EU partners from industry, city administration, universities and living labs, including Manchester City Council, IBM, Fraunhofer Institute FKIE, ENoLL, Deloitte Consulting and the National Technical University of Athens, will join efforts in the next two years, headed by IBBT, Brussels (Ballon, 2010).

PEOPLE: pilot smart urban ecosystems leveraging open innovations for promoting and enabling e-services

The PEOPLE project aims to accelerate the uptake of smart cities through an advanced deployment of innovative internet-based services in order to help them provide a better quality of life for their citizens, by applying user-driven open innovation methodologies and processes. Four pilot smart open innovation urban ecosystems (PEOPLE pilots) have been created to showcase the benefits of smarter and more sustainable cities. Establishing a social network for the PEOPLE pilots for future smart open innovation urban ecosystems and modelling activities for appraising user attributes and their identification of new service opportunities are key methodological instruments. Smart mobility and urban information management scenarios are used to develop the new e-services related to public safety information and urban living in the areas of commerce, leisure and tourism. PEOPLE's key philosophy is that social networks and integration of e-services are fundamental to building the smart city. It is suggested, therefore, that networking ICT services should be developed to enhance co-existence, focusing on various areas of the city life and the real needs of the local stakeholders.

PEOPLE uses an open data model and flows of information to develop the new internet-based services which are integrated, composed and deployed from various data sets sourced from the urban ecosystem. Four project pilots are under way: (i) Bilbao (Spain) focuses on public safety and city living's aspects of quality of life information services; (ii) Vitry sur Seine (France) deals with public safety and mobility information, particular with groups at risk of exclusion, both in terms of social life and business opportunities; (iii) Thermi (Greece) creates an 'intelligent city centre' to provide information around commerce, leisure and tourism; and (iv) Bremen (Germany) uses the technology park and the university to develop new services about campus life according to the needs of the specific users – students, researchers, visiting experts and companies. Eight service applications are being developed to sustain the identity of the area which combines innovation and commercial activities, so that innovation can develop as an entertaining activity. ANOVA IT Consulting (Italy) is leading the project; it includes eight other partners from Spain, France, Greece and Germany (Del Rozo, 2010)

Discussion

It is clear that the EU agenda is strategically moving beyond the IT paradigm of the smart city. Despite the broad and dynamic mixture of visions and approaches, however, the theme continues to attract the attention of many different sectors and professions. The most important insight coming from the analysis of current trends is that in order to meet the complex challenges of future cities, society should use the benefits of modern ICT and their full capabilities so urban communities can leverage better services and goods to the public by bringing together government, citizens and private entrepreneurs. But to bring this complex agenda forward, new and consistent smart city strategies are necessary, ones that can contribute to achieving urban sustainability and a better quality of the life for all citizens.

With the advance of technologies, society's spirit of innovation is booming. Open innovation is emerging as the new paradigm for building the smart city. With it, government and developers can draw on the expertise, skills and knowledge of the citizens to develop the advanced services and goods that are relevant to the needs of the people and the urban environment. Through open innovation the boundaries between firms, society and government open up to transfer innovation inwards and outwards in the urban ambiance and beyond, to boost research, development and delivery through partnerships and other means of facilitation. Open innovation in the smart city thus becomes part of a much broader shift that is emerging across different sectors and city networks, and most visibly between the public, private and voluntary sectors. Central to this move is a new process of co-production which understandingly calls for new models of production and consumption. Defined as 'distributed networks to sustain and manage relationships', these blur the boundaries between producers and consumers, underline systematic informal interactions and entail shared values, abilities and capacities. And with more and more networking and collaborative initiatives at local and international level, the conditions are beginning to emerge that are likely to accelerate open innovation in the future. But to move forward to a new scale, a new logic, it is necessary for principles and agendas for the smart cities to evolve.

From the present critical review of how smart cities are linked to and are dependent on open innovation, five main directions for moving forwards have emerged, and these are also a useful guide to the next steps for open innovation.

a) Raising social interaction in the heart of the smart city model, in which the infrastructures and services are jointly and dynamically discovered, invoked and composed by providers and users alike.
b) Creating open 'digital citizen-developer' communities and establishing private-public-people partnerships (PPPPs) to find dynamic and imaginative ways to interact and create, drawing inspiration and experience from open innovation and sustainable urban development.
c) Building new collaborations and networks, so cities can understand innovation, innovators understand cities, citizens can become effectively engaged and users can become content and service producers and deliverers.

d) Deploying convergent Future Internet platforms and services for the promotion of sustainable life and work styles in and across emergent networks of smart cities.

e) Creating smart open innovation urban ecosystems – specific urban settings or innovation playgrounds which combine innovation and social and commercial activities to enable open innovation and showcase the benefits for localities of growing smarter and more sustainable.

This course of development is consistent with the emerging political argument about sharing the potentials of innovation in the digital economy, despite the fact that cities articulate different open innovation approaches and smart city objectives. Notwithstanding the wide-ranging rhetoric, what is really missing in the current discussion about the smart city is the role of the territory in providing the opportunity and resources for the flourishing of open innovation. As many non-technological studies and policy initiatives have shown, the physical amenities of place, its people and their culture and environments matter a great deal for sustainable cities. This means that open innovation too has a geographical locus. The present analysis has clearly shown that open innovation in the smart city is strongly embedded locally in spite of taking the advantages of networking with other cities and communities. Therefore, capturing the true values of the territory and its total capacity – socio-economic, environmental, cultural and technological – in a manner which is consciously and strategically geared towards improving urban sustainability, governance and the urban quality of life –stands out as the greatest challenge ahead. In this line of thought, one working definition that comes out of the present review is that *open innovation in smart cities means using ICT for delivering more sustainable and inclusive cities with better quality of life for their citizens through delivering better services and goods in a mutual and creative relationship between local officials, professionals and the people, supported by the right set of strategic policies.* This implies that open innovation is not only a mindful, but also strategically driven collaboration between the stakeholders, which leads to a systematic change in the smart growth of cities.

A number of agents are key to such transformations. Human capital is at the heart of the process. Whether people are currently defined as users, clients or citizens, they all provide the vital ingredients that allow innovation to flourish and to be more effective. As urban challenges continue to grow, it seems perfectly logical to ask people for their help and to use their capacities in building their smarter city. Open innovation can offer the opportunity to transform the dynamic of the smart city by pooling the many types of skills and knowledge of the people, based on their lived experience and professional learning. This understanding is central for the idea of open innovation in the smart city. This implies, however, that the production and delivery of services and goods should be turned inside out, so cities can truly rediscover the potentials of the people and re-invigorate the social networks of

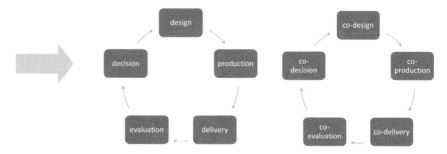

Figure 7.3 The transition of urban service production towards a sustainable open innovation model

their communities. Because open innovation goes beyond the idea of citizens' engagement in urban affairs and implies fostering new principles of mutual partnership, not just public and private, but this time public-private-people partnerships, where perhaps for the first time people can be recognised as assets and the work undertaken to sustain the city and make it sustainable and more socially just can be valued.

It is obvious that what these ambitions have in common is the shift in attitude to the users of urban services and the potentials of the citizenry in their localities to bring change. Citizens have the potential to become an effective instrument for implementing open innovation not just locally but in collaboration with other global actors, technologies and platforms. The outcome of this, where it happens, can be a radical shift of the nature and focus of the city's services, for example – not just looking at the needs of the people, but increasingly looking outwards to the potentials of the people in their concrete territory – the local neighbourhood or street, the city centre or park, the urban government or the larger region. This analysis has clearly shown that open innovation provides the option to rejuvenate the local potential of the territory, where citizens are consciously aware of and make the choice to share responsibility for their impact in a process of co-living, co-production and co-decision. It is thus the concept of the territory with its assets – physical, human, cultural and environmental – combined with the paradigms of sustainability, governance and strategic planning which is likely to drive future developments in the area.

Another important insight of this review is the transformation of the traditional model of innovation within public service provision – consisting generally of design, production, delivery and, ideally, evaluation – towards an 'outward' looking service development, where the appropriateness and the sustainability of urban services is ensured by the people's involvement in all stages of the process (see Figure7.3).

The living lab can provide the natural ecosystem for the processes of open innovation. By merging research and innovation processes with the daily, local, real-life context, close to people in their role as both citizens and consumers, using the living lab approach can enable the smart city to become a real nexus of open innovation, enabling it to tackle issues of behavioural change and innovation, building appropriate business models, organisational processes and structures, multi-stakeholder participation, and taking into account (multi-)cultural specifics. It can also broaden the scope and role of open innovation within the EU smart city agenda, providing an essential link between sustainability-oriented urban growth and the competitiveness-driven ICT development, reaching towards the collective goal of territorial cohesion and integration.

Yet practice so far has centred primarily on methods, processes and products. Participatory services and co-production of new research in the search of new ways of engaging with the users and the social networks are the latest foci. But it is also clear that a new dimension of open innovation is vital for the smart city, this being the importance of open innovation for strategic policy. Whilst their advance is admirable, living labs have not themselves developed better policy governance for the simple yet obvious fact that – to date – they have not included the citizens in the co-creation of policy. Whilst they helped encourage greater interaction between citizens and civil servants, they have not yet delivered the improved governance of location-specific and context-based policies for public services. Instead, under the existing living lab model, civil servants engage with citizens much like the private sector engages with customers – i.e. as end users to provide input into largely pre-determined new product concepts and designs or, in the case of the public sector, service models. What has been missing in this traditional equation is the proactive engagement of the users, as citizens, in the shaping and creation of the initial policy direction that ultimately determines service priorities. As the living labs innovation instrument matures it is paramount to ensure this policy dimension is further developed.

Conclusion

Building on cross-fertilisation of state-of-the-art concepts and approaches and on-going activities and programmes in Europe, the main goal of this chapter has been to explore the perspectives lying behind the question of whether the smart city is a nexus of open innovation. To find answers, the study has looked at smart cities as the enabler of social interaction, service transformation, governance and territorial rejuvenation, as the means for cities to become smarter, yet more sustainable, inclusive and attractive to people and businesses. There is considerable evidence to suggest that Europe is creating a smart city approach which has become open innovation-driven, but has yet to develop the levers it needs to enable more effective implementation of the strategies. Current trends show that smart cities initiatives are designed to foster Future Internet development and the growth of the living labs to tackle issues of ICT service and product enhancement. Given the current diversity of uses, this paper also explains that in the context of a smart city open innovation system, this means co-production and co-delivery of not just goods and services, but polices as well. The emerging insights help to increase the understanding of smart cities as nodes for open innovation and how this can itself be applied to build the smart-er city, that is more inclusive, with smarter citizens and providing better urban services.

All things said, this chapter reveals that smart cities and open innovation are dynamically interrelated – the former provides the necessary ecosystem; while the latter provides a platform for the city to become smarter. It has also been shown why, if properly understood and systematically applied, open innovation is set to create a radical shift in building more sustainable cities, which ultimately translates into better urban governance and territorial resilience. But it has also gone some way to highlight a more fundamental problem which is barely addressed in current debates – the absence of any discernible policies for building the smart city, in part because of the lack of radically new approaches inherent in existing living labs practices. With most efforts focusing on smart cities, activities proceed without understanding the key constructs and principles, and in light of this, it is clear that open innovation is something which may be stalled if a new and cohesive policy approach is not built. Collaboration too needs to be reorganised in ways that are more effective and lasting, yet the models and the policies required to achieve this are still not commonly available.

At the level of the EU, understanding and addressing open innovation strategically means embracing a common definition of open innovation in the context of smart cities, defining the strategy, mainstreaming the policy integration and identifying the areas of urban development where it can have the most profound effect. Embracing the culture of open innovation and the means for utilising the appropriate measures is equally important. But building such capacities is paramount and means establishing new communities of city officials, active citizens and professionals, whose aspirations for smarter and more sustainable cities are matched only by their grasp of the strategic developments, practical skills and knowledge needed to achieve them.

References

Amsterdam Smart City: http://www.amsterdamsmartcity.nl/#/en
Andersen, C. (2006) *The Long Tail: How Endless Choice is Creating Unlimited Demand*, London, Random House Business Books.
Ballon, P. (2010) Introducing EPIC: European Platform for Intelligent Cities, Presentation made at the Smart Cities Cluster Meeting, Brussels, 15 September.
Ballon, P., Pierson, J. and Delaere, S. (2007) Fostering Innovation in Networked Communications: Test and Experimentation Platforms for Broadband Systems., in Heilesen, S.B. and Siggaard, S., eds, *Designing for Networked Communications: Strategies and Development*, Hershey, PA/London, IDEA.

Berry, C.R. and Glaeser, E.L. (2005) The Divergence of Human Capital Levels Across Cities, *Regional Science*, 84 (3): 407–444.

Boyle, D. and Harris, M. (2009) The Challenge of Co-production: How Equal Partnerships Between Professionals and the Public are Crucial to Improving Public Services, *New Economic Foundation (NEF)*, The Lab, London, NESTA.

Cahn, E. (2001) *No More Throwaway People: The Co-production Imperative*, Washington, DC, Essential Books.

Caragliu, A., Del Bo, C. and Nijkamp, P. (2009) Smart Cities in Europe, *Serie Research Memoranda 0048*, VU University Amsterdam, Faculty of Economics, Business Administration and Econometrics, http://ideas.repec.org/p/dgr/vuarem/2009-48.html.

Carter, D. (2010) SMARTiP: Smart Metropolitan Areas Realised Through Innovation and People, http://ec.europa.eu/information_society/activities/livinglabs/docs/smartip_pub.pdf.

Chesbrough, H.W. (2003) Open Innovation: The New Imperative for Creating and Profiting from Technology, Boston, Harvard Business School Press.

Deakin, M (2007) From City of Bits to E-topia: Taking the Thesis on Digitally-inclusive Regeneration Full Circle, *Journal of Urban Technology*, 14 (3): 131–143.

Deakin, M. (2009) The IntelCities Community of Practice: The eGov Services Model for Socially Inclusive and Participatory Urban Regeneration Programs, in Reddick, C., ed., *Handbook of Research on e-Government*, Hershey, IGI.

Deakin, M. (2010) SCRAN's Development of a Trans-national Comparator for the Standardisation of eGovernment Services, in Reddick, C., ed., *Comparative E-government: An Examination of E-Government Across Countries*, Berlin, Springer Press.

Deakin, M. and Allwinkle, S. (2007) Urban Regeneration and Sustainable Communities: The Role of Networks, Innovation and Creativity in Building Successful Partnerships, *Journal of Urban Technology*, 14 (1): 77–91.

Deakin, M., Lombardi, P. and Cooper, I. (2011) The IntelCities CoP for the Capacity-building, Co-design, Monitoring and Evaluation of eGov Services, *Journal of Urban Technology*, 18 (2): 17–38.

Del Rozo, P. (2010) People Project, CIP Smart Cities Pilot presentation at Helsinki International Conference, 16 November, www.slideshare.net/openlivinglabs.

EC (2010a) i2010 – A European Information Society for Growth and Employment, http://ec.europa.eu/information_society/eeurope/i2010/index_en.htm.

EC (2010b) Europe 2020, Priorités, http://ec.europa.eu/europe2020/priorities/smart-growth/index_en.htm.

EC (2010c) Europe 2020 Strategy – Innovation Insights from European Research in Socio-economic Sciences, 1 June 2010, Brussels, Belgium, http://ec.europa.eu/research/social-sciences/events-107_en.html.

EC (2010d) Living Labs for User-driven Open Innovation, Directorate General for the Information Society and Media, http://ec.europa.eu/information_society/activities/livinglabs/index_en.htm.

EC (2010e) Smart Cities and Future Internet Experimentation, Future Internet Assembly, Ghent, December, Session VI, Smart Cities, www.future-internet.eu/home/future-internet-assembly/ghent-dec-2010/session-vi-smart-cities.html.

EC (2010f) *Towards a Future Internet*, Brussels, DG information Society and Media.

ENoLL (2010) European Network of Living Labs, www.openlivinglabs.eu/aboutus.

Europe Future Internet Portal (2010) www.future-internet.eu/home/future-internet-assembly/ghent-dec-2010/session-vi-smart-cities.html.

Europeasmartcities (2011) http://smart-cities.eu/why-smart-cities.html.

Glaeser, E.L. and Berry, C.R. (2006) *Why are Smart Places Getting Smarter?*, Taubman Center Policy Brief, Cambridge, MA, Taubman Centre, 2006–2.

Graham, S. and Marvin, S. (1996) *Telecommunications and the City*, London, Routledge.

Hollands, R.G. (2008) Will the Real Smart City Please Stand Up?, *City*,12 (3): 303–320.

IntelCities, www.intelcitiesproject.com.

Jensen, S., eds (2007) *Designing for Networked Communications*: *Strategies and Development*, Hershey, PA/London, IDEA.

King, S. and Cotterill, S. (2007*)* Transformational Government? The Role of Information Technology in Delivering Citizen-centric Local Public Services, *Local Government Studies*, 33 (3): 333–354.

Komninos, N. (2002) *Intelligent Cities: Innovation, Knowledge Systems and Digital Spaces*, London, Spon Press.

Komninos, N. (2009) Intelligent Cities: Towards Interactive and Global Innovation Environments, *International Journal of Innovation and Regional Development* (Interscience Publishers), 1 (4): 337–355 (19).

Lemke, M. and Luotonen, M. (2009) Open Innovation for Future Internet-enabled Services in 'Smart Cities, Discussion Paper Draft 2, European Commission, INFSO-F4.

Lombardi, P., Cooper, I., Paskaleva, K. and Deakin, M. (2009) The Challenge of Designing User-centric E-services: European Dimensions, in Riddeck, C., ed., *Research Strategies for eGovernment Service Adoption*, Hershey, Idea Group Publishing.

Mellander, C. and Florida, R. (2009) Creativity, Talent, and Regional Wages in Sweden, *The Annals of Regional Science*, online publication date: 15 Dec 2009.

Mitchell, W. (2005) *Placing Words: Symbols, Space, and the City*, Cambridge, MA, MIT Press.

Mossberger, K. Tolbert, K. and McNeal, R. (2008) *Digital Citizenship: The Internet, Society, and Participation*, Cambridge, MA, MIT Press.

Nijkamp, P. (2008). E pluribus unum, *research memorandum, Faculty of Economics*, Amsterdam, VU University Amsterdam.

Odendal, N. (2003). Information and Communication Technology and Local Governance: Understanding the Difference between Cities in Developed and Emerging Economies, *Computers, Environment and Urban Systems*, 27 (6): 585–607.

OECD – EUROSTAT (2005) *Oslo Manual*, Paris, OECD – Statistical Office of the European Communities.

Paskaleva, K. (2009) Enabling the Smart City: The Progress of E-city Governance in Europe, *International Journal of Innovation and Regional Development*, 1 (4): 405–422.

PEOPLE (2010) www.people-pilot.eu

Periphèria (2010) CIP-ICT PSP Call 4 2010 Pilot B: SMARTiP, Description of Work, confidential document.

Rappaport, J. (2009) The Increasing Importance of Quality of Life, *Journal of Economic Geography*, 9 (6): 779–804.

Schuller, T., Baron, S. and Field, J. (2000) Social Capital: A Review and Critique, in Baron, S., Field, J. and Schuller, T., eds, *Social Capital: Critical Perspectives*, Oxford, Oxford University Press.

Shapiro, J. (2006) Smart Cities: Quality of Life, Productivity, and the Growth Effects of Human Capital, *Review of Economics and Statistics*, 88 (2): 324–335.

SmartCitiesProject (2011) http://www.smartcities.info/aim

SMARTiP (2010) CIP-ICT PSP Call 4 2010 Pilot B: SMARTiP, Description of Work, confidential document.

Southampton City Council (2006) Southampton Smartcities Card, www.southampton.gov.uk/living/smartcities/. Retrieved 12 November 2009.

UK Cabinet Office (2008) *Show us a better way*, http://webarchive.nationalarchives.gov.uk/20100807004350/http://showusabetterway.co.uk.

UK Ministry of Justice (2008) *Building Democracy*, www.justice.gov.uk/news/newsrelease300708a.htm.

Wikipedia (2011) Smart city definition, http://en.wikipedia.org/wiki/Smart_city.

Williamson, D. and Alperovitz, G. (2003) *Making a Place for Community*, New York, Routledge.

Winters, J. (2010) Why Are Smart Cities Growing? Who Moves and Who Stays?, *Journal of Regional Science*, no. doi: 10.1111/j.1467-9787.2010.00693.x

Wise, E. and Høgenhaven, C. (2008) *User-driven Innovation*, Oslo, Research Policy.

8   The triple helix model of smart cities

A neo-evolutionary perspective[1]

Mark Deakin and Loet Leydesdorff

Introduction

This chapter sets out to demonstrate how the triple helix model enables us to study the knowledge base of an urban economy in terms of civil society's support for the evolution of cities as key components of innovation systems. We argue that cities can be considered as densities in networks among three relevant sub-dynamics: the intellectual capital of universities, the industry of wealth creation and their participation in the democratic government of civil society. It goes on to suggest that the effects of these interactions generate dynamic spaces within cities where knowledge can be exploited to bootstrap the technology of innovation systems. In particular, spaces within cities that bootstrap the all-pervasive technologies of information-based communications (ICTs) to exploit the potential that such bottom-up reinventions have to be not only intelligent but smart in (re)generating the knowledge base of their regional innovation systems.

The innovation of creative, intelligent and smart cities

Over the past decade cities have increasingly become the object of academic interest, and of scientific and technical study. Perhaps the most noticeable exponents of this rise in academic interest can be seen in the work of Landry (2008), Komninos (2008) and Hollands (2008), on the innovation of creative, intelligent and smart cities. Collectively they serve to highlight some of the most pressing socio-demographic issues currently facing the scientific and technical community: the need for cities to be(come) innovative hubs and creative milieus and the requirement for their institutions to be not only intelligent but smart. Together they do much to set out the institutional learning and the scientific and technical knowledge base of such developments. Separately they also offer a series of critical insights into the limitations of such representations and of the notions of cities as creative, intelligent and smart.

Nowhere is this better illustrated than in the notion of the smart cities promoted by leading transnational organisations like IBM, Cisco, Siemens and E.on, and now increasingly being endorsed by industry and governments across the world. For in popularising this notion of the city, the scientific and technical knowledge base of industry and government are not only now expected to be creative, or intelligent, but smart in instrumentalising the world by way of ICTs and through the cultivation of such applications within selected environments.

While such a corporate governance-driven perspective of smart cities is a development that has already been subject to criticism by Hollands (2008), this chapter seeks to get beyond the insights that this criticism offers. Moreover, it proposes to do this by suggesting:

- It is the institutionalisation of university-industry-government relations that is critical to this development, as well as the ability that ICTs have to act as a reflexive layer in the reinvention of cities, not as merely creative, or intelligent, but smart.
- This reinvention of cities as smart is also what lies behind the surge of academic interest currently being directed at communities as the practical manifestations of intellectual capital and the knowledge this produces.
- These communities of practice in turn produce a knowledge of smart cities that cannot be defined as a top-level Mode 2 trans-disciplinary issue, without a great deal of bottom-up cultural reconstruction.

Taking such a bottom-line approach to the reinvention of cities, the chapter serves to kick-start this cultural reconstruction. This is done by challenging the Mode 2 assumption that such development is the spontaneous product of market economies, and by using the critical insights that the triple helix model offers to represent the policies, academic leadership qualities and corporate strategies that provide critical insights into the governance of this cultural reconstruction. This reveals that cultural development, however liberal and potentially free, is not a spontaneous product of market economies, but a product of the knowledge-based policies which need to be discursively and carefully constructed. Otherwise, cultural development of this kind remains merely a series of symbolic events, left without the analytical frameworks needed to explain itself in terms of anything but the requirements of the market. This perspective also serves to demonstrate that any such appeal to the efficiency of the market as a means to explain cultural development can only be considered as mere analytical shortcuts, holding back any meaningful specification of the policies that their governance calls for.

Drawing upon the renaissance experiences of world-class cities like Montreal and Edinburgh, the chapter provides evidence to show how entrepreneurship-based and market-dependent representations of knowledge production are being replaced with just such a community of policy makers, academic leaders, corporate strategies and alliances. Policies, leaders, strategies and alliances with the potential to liberate cities from the stagnation they have previously been locked into, and to offer communities the means to reach beyond the idea of creative slack. Beyond the idea of creative slack and towards a process of reinvention whereby cities become 'smarter', in using intellectual capital not only to meet the efficiency requirements of wealth creation, but also to become centres of creative slack, distinguished by virtue of their communities being not only economically innovative, or culturally creative, but enterprising in opening up, reflexively absorbing and discursively shaping the governmental dimensions of such developments.

Armed with these critical insights, the neo-evolutionary perspective of the triple helix model is subsequently used as a means to uncover the intellectual capital sustaining the development of this cultural reconstruction and to reveal how it is possible for this process of reinvention to function as a set of meta-stabilising mechanisms for integrating cities into the emerging innovation systems.

In using the triple helix model to provide these critical insights into the evolution of smart cities, the chapter subdivides into four sections. The first examines the shift from the so-called Mode 2 to triple helix accounts of emerging innovation systems. The second draws upon the triple helix to account for the ongoing reconstruction of Montreal and Edinburgh as smart cities. The third outlines the critical distinction between Mode 2 and triple helix accounts of knowledge production in smart cities. The examination then goes on to reflect on the critical role this cultural reconstruction of cities, i.e. as smart, takes within regional innovation systems.

From Mode 2 to triple helix accounts of emerging innovation systems

The proponents of the Mode 2 thesis argue that the social system has undergone a radical transformation and this has changed the prevailing mode of knowledge production. Advocates of the Mode 2 thesis argue that disciplinary-based knowledge shall increasingly become obsolete and will be replaced with techno-scientific knowledge generated in trans-disciplinary projects. Within these arguments, the concept of national *systems* of innovations, as it prevails in evolutionary economics, focuses on the resilience of existing arrangements – which are in the process of being creatively destructed (Schumpeter, [1939], 1964).

Extensive research carried out in evolutionary economics has enabled systematic comparisons of different innovation systems without much questioning of whether nations are the level at which innovation systems are integrated (Lundvall, 1992; Nelson and Winter, 1982; Nelson, 1993). In addition to the idea of the nation state – as a specific construct of the nineteenth and twentieth centuries – providing a stable context for the development of *national* innovation systems, other scholars have sought to focus on the emergence of *sectorial* or *regional* systems as potential candidates for the stabilisation of interactions among selection environments (Braczyk et al., 1998; Carlsson, 2006; Carlsson and Stankiewicz, 1991).The triple helix, on the other hand, explains these differences among innovation systems at different levels in terms of *possible* arrangements. For example, when the nations of Eastern Europe became transition economies after the demise of the Soviet

Union in 1991, the ambitions of these countries to develop national systems of innovation met with interference from market forces, on the one hand, and from the ongoing political process of European accession, on the other. An interesting example of how this worked is provided by the case of Hungary, where not one but three innovation systems emerged during the transition (Inzelt, 2004).

Here a metropolitan centre developed around Budapest to compete with cities such as Vienna, Munich and Prague as a seat for knowledge-intensive services, multinational corporations, etc. In the western part of the country, specific Western European companies also moved into Hungary to the extent that they were able to influence research agendas at universities. The German car manufacturer Audi set up its own institute at a local university in the north-western region where it developed an automotive cluster (Lengyel et al., 2006). In addition to this, a third type of innovation system could be detected in the eastern regions, where traditional universities support the development of local infrastructures, remaining more continuous with the old system (Lengyel and Leydesdorff, 2011).

These results indicate that when Hungary arrived on the European scene, it was too late to develop a purely *national* innovation system because the envisaged system was already implicated in the formation of the European Union. Transition countries became at the same time accession countries for the European Union and the resulting dynamics could henceforth only be coordinated loosely at the national level. The period for adaptation was too short for stabilising a national system of innovations. The context of the European Union has changed the status of regions: nation states can be dissolved as in the case of Czechoslovakia, or continuously reformed as in the case of Belgium.

Under the knowledge-based regime, each system can be expected to remain in endless transition (Etzkowitz and Leydesdorff, 1998). However, this endless transition does not mean that anything goes, but rather a continuous recombination of strengths and competitive advantages under selection pressures (Cooke and Leydesdorff, 2006). The selection processes involved are knowledge intensive because they can only be improved by appreciating the information that comes available when they operate.

Such disorganising tendencies may vary from country to country and from region to region within countries. In the case of Eastern Europe, the transition not only represents a newly emerging trajectory, but a change in the regime regulating all of this. This emerging system, however, should not be reified as another Mode 2 system: the interacting *uncertainties* in the distributions determine the dynamics by selecting upon one another. One can no longer expect a stable centre where decision making can be monopolised because the one-to-one correspondence between functions and institutions no longer prevails. Under these circumstances, the fragile order of the prevailing knowledge base remains a networked order of codified expectations.

The disorganisation and fragmentation of previously existing innovation systems is appreciated in the triple helix model in terms of a reflexive overlay of relations among the carriers of innovation systems (Etzkowitz and Leydesdorff, 2000). Here the overlay feeds back as a restructuring sub-dynamic on the underlying networks, and generates and/or blocks/outcompetes opportunities for either sectoral or regional niche-formation in a distributed mode. New competencies may be needed for further developments and new specialities are shaped as a recombination of existing disciplinary capacities. The knowledge-based dynamics are institutionally conditioned, but evolutionary in character: the reflection at the level of the overlay operates from the perspective of hindsight and can therefore be future oriented. These dynamics generate flexibilities; not as a biological process of adaptation, but as a *social* dynamic of interactions among meanings, insights and intentions (Freeman and Perez, 1998; Leydesdorff, 2009, 2010).Unlike the national systems and Mode 2 accounts of knowledge production, the triple helix model:

- studies networks of university-industry-government relations and offers a neo-evolutionary model of a knowledge-based economy;
- proposes that the three evolutionary functions shaping the selection environments of a knowledge-based economy are: (i) organised knowledge production, (ii) wealth creation and (iii) reflexive control;
- suggests that as reflexivity is always involved as one of them, the functions that they serve are not given, but socially constructed as the inter-human co-ordination mechanisms of evolving communication systems within given cultural settings.

In the triple helix model the dynamics of these social construction and inter-human co-ordination mechanisms are endogenous, because actors in the three institutional spheres relate to each other reflexively. Integration and differentiation among the subsystems are concomitant: the functionally differentiated system is able to process more complexity, while (integrating) exchange relations among the subsystems make it possible to change perspectives and to develop historically new structures at interfaces. Here one can expect a configuration to be reproduced in which the generation of intellectual capital prevails within an academic environment, with wealth creation being institutionally associated with industry, while control in the public sphere can be associated with government. On the other hand, network relations can be expected to reflect degrees of integration; for example, in national systems. The degree of integration and whether synergy is generated, however, remain empirical questions which are open to measurement (e.g. Leydesdorff and Fritsch, 2006; Leydesdorff and Sun, 2009).

Under this model institutions are seen as reacting to each other's selections (Etzkowitz, 2008). The dynamic of this selection process is not biologically inherited (Lewontin, 2000), but cultural, i.e. dependent on the development of communicative competencies – i.e. learning – by the carrying agents. The interactions among the dynamics in the overlay can be intensified by the technologies of information-based communications (ICTs) currently being exploited to generate the notion of creative cities (Landry, 2008) and as the knowledge base of intelligent cities (Komninos, 2008). Technologies that Hollands (2008) notes are now being asked to become even smart-er. Smarter, not just by the way they make it possible for cities to be intelligent (as an institutional agent) in accumulating capital and creating wealth, but in entertaining their governance.

Such a co-evolutionary mechanism for the meta-stabilisation of existing institutional arrangements marks a development that takes us beyond the dismantling of national systems and construction of regional advantages, i.e. accounts that fall under the remit of innovations systems and Mode 2. We suggest the reinvention of cities currently taking place under the so-called urban renaissance cannot be defined as a top-level trans-disciplinary issue without a considerable amount of cultural reconstruction. Although recognised as important by advocates of the Mode 2 perspective, the highly distributed and local character of this reconstruction has to be appreciated as a driver of the transformations.

In our opinion, accounts of this cultural reconstruction have tended to reify the global perspective and fail to appreciate the meta-stabilising transformation of innovations systematically worked out as the informational content of social and cultural processes operating at the local level. We also suggest, the potential this meta-stabilising mechanism has to work as a reflexive layer is what lies behind the surge of academic interest currently focusing on communities as the practical manifestation of intellectual capital and as the means to exploit the knowledge generated from their organisation (Amin and Cohendet, 2004; Amin and Roberts, 2008).

The triple helix, however, also distinguishes between the codes of communication operating within these communities of practice (CoPs) and highlights the need to specify translation mechanisms among them (Nooteboom, 2008). We suggest that the translations between and among such communications can generate intellectual capital and provide new sources of material for this meta-stabilising dynamic. For innovation systems can use this knowledge base as a means to counter stagnation and draw upon their networks to develop another – that is, analytically orthogonal, or third – selection mechanism, which operates between and upon the market forces of institutionally oriented policies. From this perspective, the transformation of national systems to the types of conceptual frames outlined in the rest of this chapter offers an opportunity for the nation to be considered as one among a number of competing hypotheses.

The ongoing reconstructions of Montreal and Edinburgh

We suggest under the conceptual frames set out in this chapter, the decisive scale of territorial organisation is often not national but global. For example, cities like Montreal and Edinburgh, although not capital cities, in a national innovation system can obtain the status of world class as transnational city-regions that function as innovation systems. Montreal, for example, has been recognised as a city that is particularly successful in reinventing itself and developing a 'creative' force within the region (Florida, 2004; Stolarick and Florida, 2006). While informal communities are found to generate new knowledge, the city has sought to institutionalise this process of production by developing into a learning organisation. This organisational structure has in turn invented pedagogy by which to integrate knowledge-intensive firms into the metropolitan innovation system. Furthermore, this pedagogy has then developed the means to integrate and exploit them as the key components of the emerging innovation system.

As an empirical example, Figure 8.1 shows the long-term trends in the ratios of patents for Edinburgh versus Glasgow (as a sister city), measured on 15 February 2011, in the database of the World Intellectual Property Organisation (WIPO) in Geneva. We juxtaposed these results with similar data for Montreal versus Calgary as two provincial metropoles in Canada, but in this case using USPTO data. Although the industrial and traditional resource bases of both Calgary and Glasgow are much stronger, the patenting in the two smarter cities (Montreal and Edinburgh) is slowly gaining ground because of the ongoing transformation from a political economy (organised around the notion of a nation state) to a (globalised) knowledge-based economy.

As Cohendet and Simon (2008) have noted, under these conditions it is not just universities, industry or governments but communities that provide the environments by which it becomes possible for cities to successfully exploit the opportunity to manage such integration. Exploit it up to the point where the city of Montreal has learnt how to become a leading exponent of cultural events and known for the advantage such an innovation system manages to construct (Nowotny, 2008). In this case, it might be said, the flow of cultural events into and out of intellectual capital, wealth creation and the government of civil society interacts to open up new windows and perspectives.

The only thing offered to explain the growth of Montreal as a leading exponent of cultural events from this institutional perspective has, however, hitherto been a list of enabling conditions, such as a strong research, scientific and technical development ethic, whose shared enterprises are underpinned by leading university involvement. By leading university involvement which is in turn supported by strong leadership from the city and a set of policies which are capable of governing such ventures as part of an urban regeneration programme.

In contrast to this, Edinburgh's strategy has been to expound the virtuous nature of smart cities, not on the basis of leading university involvement, supported by strong leadership, but from the application of ICTs to develop electronically enhanced services capable of meeting their evolving e-government agenda (Malina and Macintosh, 2004). Here an attempt has been made to get beyond policy statements about the enabling conditions of strong research,

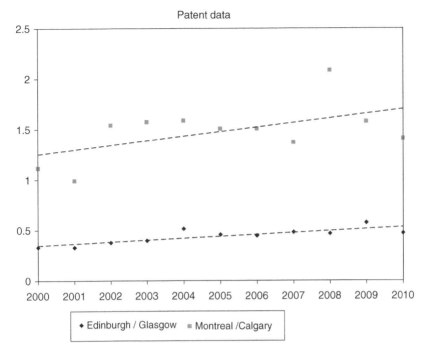

Figure 8.1 Long-term trends in ratios of patents between Edinburgh versus Glasgow (WIPO data) and Montreal versus Calgary (USPTO data)

scientific and technical development commitments and towards a series of actions capable of meeting the governance requirements of urban regeneration programmes (Malina and Ball, 2005).

As with Montreal, the actions being promoted by Edinburgh have all highlighted the significance of community. In contrast to Montreal, this sense of community has developed into the IntelCities community of practice and integration of the eGovernment services this develops to meet the requirements of Edinburgh. At present this integration is mainly technical, concerning the software developments needed to host such services and meet with the semantics of the platform's e-learning needs, knowledge transfer requirements and capacity building commitments. This currently takes the form of an 'eTopia' demonstrator, showing in 'session-managed logic' how the eCity platform accesses the extensive pool of eGov services located in the back office and uses the intelligence embedded in the middleware to deliver Level 3 (advanced e-Citizenship) courses on the consultative needs and deliberative requirements of such developments. This provides a real-time demonstration of the platform's capacity for Edinburgh to be smart in developing the technical and semantically rich content required for the middleware to begin supporting the socially inclusive consultations and participatory deliberations of urban regeneration programmes.

These enhanced processes of consultation and deliberation also have the advantage of offering citizens multi-channel access to such eGov services, customised, co-designed and bundled together as socially inclusive and participatory urban regeneration programmes for bringing about improvements in the quality of life (Anthopoulos and Fitsilis, 2010; Anthopoulos and Vakali, 2012). This goes a long way to:

- uncover the business logic needed to base the intelligence-driven (re)organisation of cities on, and the standards required to benchmark the performance of the platform against;
- provide the performance-based measures needed to assess whether the plans cities posses to develop eGov services (over the platform) have the embedded intelligence (the learning, knowledge-based competencies and skills) required to support such actions;
- also provide the means to evaluate if such planned developments build the (intellectual) capacity – learning, knowledge-based competencies and skills – needed to support such actions.

In addition to developing the semantically interoperable eGov services, Edinburgh's smart city strategy has also sought to evaluate how well these have been bootstrapped as components of the eCity platform. This has meant developing three eTopia demonstrator storylines, where the typical learners use the eCity platform to query the development of urban regeneration programmes by either searching for information on a given initiative, gaining access to possible online transactions supporting any such actions, or by getting involved in the customisation and co-design of those consultations and deliberations underlying the governance of such proposals. These three storylines developed scenarios for: 1) accessing local services in neighbourhoods subject to regeneration; 2) carrying out online transactions related to the use of land; 3) consultations and deliberations about the safety and security issues underlying the governance of urban regeneration programmes.

As can be appreciated, any reduction of these interaction effects among the intellectual capital and wealth creation of their respective governance regimes, to one contextualizing the other as mere conditions, tends to lead the discourse on smart cities towards an economic representation of innovation systems, or a singularly one-sided account of their scientific and technical qualities from the trans-disciplinary perspective of Mode 2 knowledge production (Hessels and van Lente, 2008). In the former the discourse tends to be economic (as illustrated in the case of Montreal), whereas with the latter (as with the situation in Edinburgh) the managerial perspective prevails.

What both have in common, though, is the tendency for such discourses to take on forms that are unable to uncover the socially organised learning processes which the knowledge base of smart cities is understood to cultivate.

This is because with these discourses the products of science and technology are absorbed as exogenous fruits from heaven, which in turn means the process of absorption that is fostered by way of their associated polices, leadership and corporate strategies cannot be improved other than by adaptation and imitation.

From our neo-evolutionary perspective, however, these predominantly social situations and the communities they give rise to can be cultivated as relevant selection environments. The codes operating in these selection environments can be reconstructed, adjusted and sometimes strengthened by interacting in local settings. The interactions can be improved by learning how to translate the communication from one context into another.

The critical distinction

We suggest that the critical distinction between the Mode 2 type accounts of creative communities, set out by Florida (2004) and Stolarick and Florida (2006) in relation to Montreal, and the likes of Malina and Macintosh (2004), Malina and Ball (2005), Anthopoulos and Fitsilis (2010) and Anthopoulos and Vakali (2012) for Edinburgh and those of the triple helix model, lies in the tendency for:

- the former to remain managerial and become locked into neo-liberal policies displaying a strong entrepreneurial legacy, and then to be articulated with reference to the market economy and its regime of accumulation;
- the latter to provide a framework for analysis capable of elaborating on what the intellectual capital of universities, wealth creation of industry and governance of civil society can each contribute to the knowledge-intensive policies, academic leadership and corporate strategies of emerging innovation systems.

Knowledge-intensive polices have to be articulated before they can be exploited through the scientific management of the corporate strategies governing civil society's experience of such developments. Any entrepreneurial drive to by-pass the articulation of these knowledge-intensive polices fails to represent the intellectual capital, wealth creation and governance invested in cultural construction (Ramaprasad and Sridhar, 2010).

Using the triple helix, it can be recognised that cultural development, however liberal and potentially free, is not a spontaneous product of market economies, but a product of the policies, academic leadership and corporate strategies that need to be carefully reconstructed as part of an urban regeneration programme. Otherwise, cultural development of this type risks remaining a series of symbolic events, left without the analytical frameworks needed for selection environments to legitimise themselves in terms other than economic success.

Any such appeal to the efficiency of the market as a means to explain cultural development serves as an analytical shortcut, holding back a more meaningful specification of the policies, leadership qualities and corporate strategies underpinning an urban regeneration programme. For cities like Montreal and Edinburgh show how the creative ecology of an entrepreneurship-based and market-dependent representation of knowledge-intensive firms can be replaced with a learning organisation of policy makers, academic leaders and corporate strategists. That is with a set of alliances which in turn have the potential to liberate cities from the stagnation they have previously been locked into and to offer communities the freedom to develop policies, with the leadership and strategies, capable of reaching beyond the idea of creative slack as merely a residual factor. For in order to be more than intelligent and smart, and in that sense smarter, cities need the intellectual capital required to not only meet the efficiency requirements of wealth creation under the enterprise of a market economy, but to become centres of creative slack distinguished by virtue of their communities having the political leadership and strategies that are capable of not only being culturally creative, but reflexive in absorbing and discursively shaping their governance.

This neo-evolutionary perspective on the dynamics of learning and their codification of knowledge can also guide us towards the intellectual capital of such creativity by focusing attention on those dimensions of corporate management that make it possible for cities to be smart. In particular, the intellectual capital and creativity of corporate management which makes it possible for cities to be smart in organising their urban regeneration programmes to function as meta-stabilising mechanisms that underpin civil society's integration of cities into emerging innovation systems. An account of how these learning processes and their codification of knowledge can guide us to the intellectual capital and creativity of corporate management is set out in Deakin and Allwinkle (2007) and Deakin's (2008, 2009, 2011) representation of smart cities.

Without this type of cognitive deconstruction and analysis, these representations of smart cities, their cultural events, knowledge products and intellectual achievements, would run the risk of being reified as little more than signifiers of a market economy. However, the reflexive turn set out in this chapter avoids this and allows the best practice examples to be evaluated – instead of imitated – in terms of functional advantages, and while there may be no single best practice from an evolutionary perspective, this is not critical. For as long as there is sufficient slack in the environment to learn from failures and accommodate alternatives in ways that offer the prospect of bootstrapping self-regenerating actions, this is all that matters. The different functionalities this produces can then in turn be articulated into specific policies, informed and further improved by learning from what works and uncovering the reasons why.

Such a critical approach – when compared to the boasting of self-proclaimed best practices – challenges policy makers to raise additional questions, such as whether the university should participate in such an integration, and if so how. What other potentials exist, but have hitherto been insufficiently articulated towards industry? For example, should technologically oriented faculties only be involved, or might this also include the social sciences? These questions arise because here the technology of city-regions surfaces for what they can rightly be considered to be: nested centres of control, dependent for their further economic and social development not only on the market, but also on intellectual capital and wealth creation of the reflexive and self-organizing systems this cultivates.

Some reflections

Using the triple helix model, it can be recognised that cultural development, however liberal and potentially free, is not a spontaneous product of market economics, but a product of the policies that need to be carefully constructed by a governing authority. For:

- cities like Montreal and Edinburgh show how the creative ecology of an entrepreneur-based and market dependent representation of knowledge-intensive firms is currently being replaced by a community of policy makers, academic leaders and corporate strategists;
- these communities in turn have the potential to liberate cities from the stagnation that they have previously been locked into and offer the freedom to develop polices, with the leadership and strategies capable of reaching beyond the idea of creative slack as a residual factor;
- in order for them to be more than intelligent and smart, and in that sense, smarter, cities need their intellectual capital to not only meet the efficiency requirements of wealth creation under a market economy, but to become centres of creative slack distinguished by virtue of their communities having the political leadership and strategies capable of not only being culturally creative, but enterprising in using ICT-saturated forms of knowledge-intensive production to open up, reflexively absorb and discursively shape their governance.

Conclusions

This chapter has set out to demonstrate how the triple helix model enables us to study the knowledge base of an urban economy in terms of civil society's support for the evolution of cities as key components of innovation systems. In this schema, cities can be considered as densities in networks among at least these three relevant dynamics: that is, in the intellectual capital of universities, in the industry of wealth creation and their participation in the democratic government that forms the rule of law in civil society. The effects of these interactions can generate spaces within the dynamics of cities where knowledge production may be exploited. The densities of relations that exist among the spaces of these institutional spheres in turn create a dynamic which makes it possible for cities to bootstrap the technology of their regional innovation systems.

These technologies, we have argued, are enabled by the all-pervasive technologies of information-based communications currently being exploited to generate the notion of creative cities, as the knowledge base of intelligent cities and their augmentation into smart(er) cities. Cities which are smart(er) at exploiting information and communication technologies that are not only creative, or intelligent in generating intellectual capital and creating wealth, but in the sense that the selection environments governing their knowledge production make it possible for them to become integral parts of emerging innovation systems. The specificity of possible matches is not given, but can be reconstructed, and remain reflexively accessible, knowledge-intensive and fragile due to the fact that discursive knowledge is based on representations which can be further informed.

We have suggested that it is the reflexive instability of knowledge-based systems that provides the co-evolutionary mechanism between institutional stabilisation and communicative meta-stabilisation and, as such, offers the possibility of relating the city to their next-order dynamics in a process of globalisation. The capacity to process this transition reflexively, that is, in terms of translations, marks a development which takes us beyond the dismantling of national systems and the construction of regional advantages. For in using this neo-evolutionary perspective of the triple helix, it can be appreciated that cultural development, however liberal and potentially free, is not a spontaneous product of market economies, but the outcome of policies, academic leadership qualities and corporate strategies, all of which need to be carefully reconstructed, pieced together and articulated before their governance requirements can be met.

Websites

Cisco's website on smart cities and communities: www.smartconnectedcommunities. org/index.jspa.

E.on'swebsiteonsmartcities:www.eonuk.com/downloads/The_MKSmart2020project. pdf.

IBM's website on smart cities: www-935.ibm.com/services/us/gbs/bus/html/smarter-cities.html.

Siemen's website on smart and sustainable cities: www.siemens.co.uk/en/news_press/ index/news_archive/siemens-to-deliver-innovative-smart-grid-solution.htm.

Notes

1 This is a revised version of Leydesdorff, L. and Deakin, M., The Triple Helix Model of Smart Cities: A Neo-Evolutionary Perspective, *Journal of Urban Technology*, 18 (2), 2011: 53–63.

References

Amin, A. and Cohendet, P. (2004) *Architectures of Knowledge*, Oxford, Oxford University Press.

Amin, A. and Roberts, J. (2008) Knowing in Action: Beyond Communities of Practice, *Research Policy*, 37 (2): 353–369.

Anthopoulos, L. and Fitsilis, P. (2010) From Online to Ubiquitous Cities: The Technical Transformation of Virtual Communities, *Lecture Notes of the Institute for Computer Sciences, Social Informatics and Telecommunications Engineering*, 26 (9): 360–372.

Anthopoulos, L. and Vakali, A. (2012) Urban Planning and Smart Cities: Interrelations and Reciprocities, *Lecture Notes in Computer Science*, 7281: 178–189.

Braczyk, H.J., Cooke, P. and Heidenreich, M., eds (1998) *Regional Innovation Systems*, London/Bristol, University College London Press.

Carlsson, B. (2006) Internationalization of Innovation Systems: A Survey of the Literature, *Research Policy*, 35 (1): 56–57.

Carlsson, B. and Stankiewicz, R. (1991) On the Nature, Function, and Composition of Technological Systems, *Journal of Evolutionary Economics*, 1 (2): 93–118.

Cohendet, P. and Simon, L. (2008) Knowledge-intensive Firms, Communities and Creative Cities, in A. Amin and J. Roberts, eds, *Community, Economic Creativity and Organisation*, Oxford, Oxford University Press.

Cooke, P. and Leydesdorff, L. (2006) Regional Development in the Knowledge-based Economy: The Construction of Advances, *Journal of Technology Transfer*, 3 (1): 5–15.

Deakin, M. (2008) The Search for Sustainable Communities: Ecological-integrity, Equity and the Question of Participation, in R. Vreeker, M. Deakin and S. Curwell, eds, *Sustainable Urban Development, Volume 3: The Toolkit for Assessment*, Oxford, Routledge.

Deakin, M. (2009) The IntelCities Community of Practice: The eGov Services Model for Socially-inclusive and Participatory Urban Regeneration Programmes, in C. Reddick, ed., *A Handbook of Research on Strategies for Local E-Government Adoption and Implementation: Comparative Studies*, Hershey, IGI Global.

Deakin, M. (2010) SCRAN: The Development of a Trans-National Comparator for the Standardisation of eGovernment Services, in Reddick, C. ed., *Comparative E-Government: An Examination of E-Government Across Countries*, Berlin, Springer Press.

Deakin, M. (2011) From Intelligent to Smart Cities: CoPs as Organizations for Developing Integrated Models of eGovernment Services, in Bulu, M., ed., *City Competitiveness and Improving Urban Subsystems: Technologies and Applications*, Hershey, ICI Global.

Deakin, M. (2012) Intelligent Cities as Smart Providers: CoPs as Organizations for Developing Integrated Models of eGovernment Services, *Innovation: The Journal of Social Research*, 14 (2): 115–136.

Deakin, M. and Allwinkle, S. (2007) Urban Regeneration and Sustainable Communities: The Role of Networks, Innovation and Creativity in Building Successful Partnerships, *Journal of Urban Technology*, 14 (1): 77–91.

Etzkowitz, H. (2008) *The Triple Helix: University-Industry-Government Innovation in Action*, Abingdon, Routledge.

Etzkowitz, H. and Leydesdorff, L. (1998) The Endless Transition: A 'Triple Helix' of University-Industry-Government Relations, Introduction to a Theme Issue, *Minerva*, 36: 203–208.

Etzkowitz, H. and Leydesdorff, L. (2000) The Dynamics of Innovation: From National Systems and 'Mode 2Mode-2' to a Triple Helix of University-Industry-Government Relations, *Research Policy*, 29 (2): 109–123.

Florida, R. (2004) *The Rise of the Creative Class: A Toolkit for Urban Innovators*, New York, Basic Books.

Freeman, C. and Perez, C. (1998) Structural Crises of Adjustment, Business Cycles and Investment Behavior, in G. Dosi, C. Freeman, R. Nelson, G. Silverberg and L. Soete, eds, *Technical Change and Economic Theory*, London, Pinter.

Hessels, L. and van Lente, H. (2008) Re-thinking New Knowledge Production: A Literature Review and Research Agenda, *Research Policy*, 37 (4): 740–760.

Hollands, R. (2008) Will the Real Smart City Stand Up?, *City*, 12 (3): 302–320.

Inzelt, A. (2004) The Evolution of University-Industry-Government Relationships during Transition, *Research Policy*, 33 (6/7): 975–995.

Komninos, N. (2008) *Intelligent Cities and Globalisation of Innovation Networks*, London, Taylor & Francis.

Landry, C. (2008) *The Creative City*, London, Earthscan.

Lengyel, B. and Leydesdorff, L. (2011) Regional Innovation Systems in Hungary: The Failing Synergy at the National Level, *Regional Studies*, 45 (5): 677–693.

Lengyel, B., Lukács, E. and Solymári, G. (2006) A külföldi érdekeltségű vállalkozások és az egyetemek kapcsolata Győrött, Miskolcon és Szegeden, *Tér és Társadalom*, 4: 127–140.

Lewontin, L. (2000) *The Triple Helix: Gene, Organism, and Environment*, Cambridge, MA/London, Harvard University Press.

Leydesdorff, L. (2009) The Non-linear Dynamics of Meaning-processing in Social Systems, *Social Science Information*, 48 (1): 5–33.

Leydesdorff, L. (2010) The Communication of Meaning and the Structuration of Expectations: Giddens' 'Structuration Theory' and Luhmann's 'Self-organization', *Journal of the American Society for Information Science and Technology*, 61 (10): 2138–2150.

Leydesdorff, L. and Fritsch, M. (2006) Measuring the Knowledge Base of Regional Innovation Systems in Germany in Terms of a Triple Helix Dynamics, *Research Policy*, 35 (10): 1538–1553.

Leydesdorff, L. and Sun, Y. (2009) National and International Dimensions of the Triple Helix in Japan: University-industry-government Versus International Co-authorship Relations, *Journal of the American Society for Information Science and Technology*, 60 (4): 778–788.

Lundvall, B.A. (1992) *National Systems of Innovation*, London, Pinter.

Malina, A. and Ball, I. (2005) ICTs and Community: Some Suggestions for Further Research in Scotland, *Journal of Community Informatics*, 1 (3): 66–83.

Malina, A. and Macintosh, A. (2004) Bridging the Digital Divide: The Development in Scotland, in Ari-Veikko Anttiroiko et al., eds, *eTransformation in Governance*, Idea Group Publishing.

Nelson, R.R. (1993) *National Innovation Systems: A Comparative Analysis*, New York, Oxford University Press.

Nelson, R.R. and Winter, S.G. (1982) *An Evolutionary Theory of Economic Change*, Cambridge, MA, Belknap Press of Harvard University Press.

Nooteboom, B. (2008) Cognitive Distance between Communities of Practice and in Firms, in A. Amin and J. Roberts, eds, *Community, Economic Creativity and Organisation*, Oxford, Oxford University Press.

Nowotny, N. (2008) *Insatiable Curiosity: Innovation in a Fragile Future*, Cambridge, MA, MIT Press.

Ramaprasad, A. and Sridhar, M.K. (2010) Empowering a State's Development of a Knowledge Society, paper presented at the 8th International Conference of the Triple Helix of University-Industry-Government Relations, Madrid.

Schumpeter, J. ([1939], 1964) *Business Cycles: A Theoretical, Historical and Statistical Analysis of Capitalist Process*, New York, McGraw-Hill.

Stolarick, K. and Florida, R. (2006) Creativity, Connectivity and Connections: The Case of Montreal, *Environment and Planning A*, 38: 1779–1817.

9 SCRAN

The network

Mark Deakin and Peter Cruickshank

Introduction

This chapter outlines the relationship between the Smart Cities (inter) Regional Academic Network (SCRAN) and the triple helix model of knowledge production. The 'step-wise' logic of SCRAN's triple helix for the SmartCities venture is then set out. This draws attention to the networking of the intellectual capital underpinning SCRAN's knowledge base and learning platform. From here the chapter goes on to set out how SCRAN's wiki is being used to match the intellectual capital and wealth creation of the SmartCities triple helix with the eGovernment services developing under the North Sea's regional innovation system.

The concept of a smart city

The concept of the smart city has been introduced as a strategic device to encompass modern urban production factors in a common framework. As such it serves to highlight two emerging trends. Firstly, the growing importance of Information and Communication Technologies (ICTs) in the development of cities, and secondly the underlying significance of social and environmental capital in the profiling of their competitiveness (Deakin, 2010 and Caragliu et al., 2009). The significance of these two assets – social and environmental capital – itself goes a long way to distinguish smart cities from their more technology-laden counterparts, drawing a clear line between what otherwise goes under the name of either digital or intelligent cities.

In seeking to highlight the underlying significance of both these asset bases, this chapter draws attention to the triple helix model of knowledge production and the wiki assembled to support the development of SCRAN as a community of practice (CoP). It draws particular attention to the university-industry-government collaborations (triple helix) underlying the Web 2.0 service-orientated architecture of this knowledge infrastructure and the deployment of such technologies as an enterprise allowing communities to learn about how to standardise eGov(ernment) services as transformative business-to-citizen applications. This serves to highlight the critical role such business-to-citizen applications play in making it possible for cities to be smart in reaching beyond the transactional logic of service provision and grasping the potential that regional innovation systems have to democratise the ongoing customisation of eGov services.[1]

In this context, CoPs cover a number of situated practices, with characteristics including shared ways of 'doing things together', 'using specific tools and other artefacts' (Wenger, 1998). Recently the 'learning-by-doing' and 'action-based' logic of this model has been extended to cover virtual organisations and online interaction (Amin and Roberts, 2008).

In defining themselves as 'virtual organisations', CoPs are commonly referred to as 'type 2' communities and draw upon lessons learnt from the practical experiences of knowledge-based networks. That is, from the assembly and subsequent deployment of such networks as virtual organisations which are set up to share best practice on the development of their infrastructures. The growing social, environmental and cultural significance of such CoPs and what they contribute to the competitiveness of cities has recently been highlighted in the work published on the IntelCities CoP (Deakin et al., 2006; Deakin, 2009). This chapter shall draw upon the valuable lessons learnt from this CoP and transfer the elements of best practice that the venture offers.

After reflecting on the communication needs and technical requirements of this venture, the chapter goes on to configure SCRAN's triple helix and set out the 'step-wise' logic of the organisation's knowledge base and learning platform. From here attention turns towards the networking of the triple helix and the wiki assembled to support the development of SCRAN as a CoP. Here attention is drawn to the university-industry-government collaborations (triple helix) underlying the Web 2.0 service-orientated architecture of this knowledge infrastructure and deployment of such technologies as an enterprise for organisations to learn about how this CoP works to standardise the transformation of eGov services. In particular how it works to standardise eGov services as transformative business-citizen applications which allow cities to be smart by integrating the business sector into the North Sea regional innovation system.

Academic networking of the SmartCities venture

Figure 9.1 draws attention to the academic network underpinning the SmartCities venture known as SCRAN. In this respect it identifies the network of academic institutions, their city partners and the specific role they take on within SCRAN. As can be seen, for Edinburgh Napier University the main object of attention is methodology, and for Mechelen (MEMORI) the object of the exercise is to partner Kortrijk in their customisation of eGov services provision. In this respect, each academic institution involved in the SmartCities venture, and their respective counterparts in the industrial and government sectors, are seen as contributing something towards the

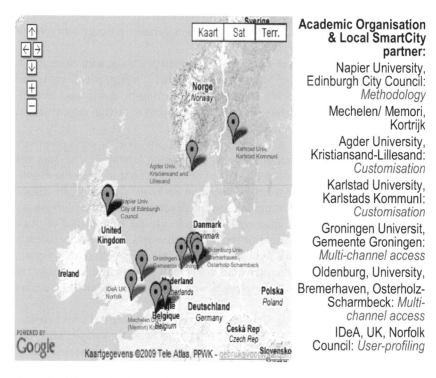

Figure 9.1 The academic organisation of SmartCities partners

knowledge base each of them needs to learn about. That is, the knowledge base that they all need to learn about as a requirement of the eGov service developments under consideration.

SCRAN as a three-way partnership

While the aforesaid draws attention to academic institutions and their city partners, it is the three-way partnership between the universities, industry and government of the network that captures the science and technology around which the 'triple helix' turns. This section offers an image of the triple helix that SCRAN proposes to develop as the three-way partnership. That is, as a three-way partnership which goes some way to capturing SCRAN's particular take on the triple helix and scientific basis of the SmartCities project, funded by the EC to support the innovative and creative use of ICTs.

From here the organisational means needed for universities and their industrial counterparts to use ICTs in the development of eGov services can be explored in terms of the said partner capabilities. That is to say, explored in terms of their ability to construct a methodology capable of not only underpining the development of eGov services, but in building the capacity to co-design them in a way that allows the monitoring and evaluation of their user profiles to be shared across the North Sea region.

The triple helix

As the main exponents of the triple helix, Etzkowitz and Leydesdorff (2002, 2003) offer a particularly insightful critique of so-called 'Mode 2' accounts of innovation, but limit their particular representation of the model to those institutional relations surrounding university, industry and government involvement in the knowledge economy of regional systems. Here attention focuses on the production of knowledge by universities and industry as an index of the intellectual capital tied up in the artefacts of innovations patented and licensed in line with the standards laid down by government to regulate such developments.

Critically insightful as these studies are, it is noticeable that many of them are limited to output-related examinations of knowledge production and tend to ignore the infrastructures that underlie the said institutional rela- tions and support the involvement of universities, industry and government in this process. Institutionally based studies of this type are limited and, as a result, little is known about either the ICTs that underlie the triple helix, or the electronically enhanced services upon which the model's understanding of knowledge production rests. Understandably, the absence of such stud- ies has led to criticism about the usefulness of the triple helix as a model of knowledge production and its value as a regional innovation system. Jensen and Bjorn (2004), Jauhiainen and Suorsa (2008) and Smith (2007) raise such concerns about the triple helix and begin to question the practical worth of the model. The apparent reluctance of the main exponents of the triple helix to engage in a debate about the model's practical worth also serves to raise doubts about its capacity to adequately account for the process of knowledge production.

SCRAN's take on the triple helix

To overcome these limitations, it is clear that SCRAN's take on the triple helix needs to develop practical guidelines on how to use the model and this requires it to cover all three strands of the helix, i.e. universities, industry and government alike. In methodological terms, the challenge this poses means SCRAN has to account for how the triple helix of SmartCities can organise the production of knowledge:

- internally (i.e. as a smart city);
- externally as part of a regional innovation system
- inwardly by representing the:

- triple helix of smart cities;
- the organisation of these helices (as the social capital of this knowledge production);
- the collaboration needed for the intellectual capital of universities and wealth creation of industry to be smart in constructing the advantage this offers cities to learn about how their communities can meet their eGov service development commitments;
- the consensus-building needed for this reinvention of cities to be smart in supporting the development of 'trans-national' standards regulating the development of eGov services;
- the practical application of such standards in building the capacity required to be post-transactional, that is to say, post-transactional in the sense in which business and citizens are able to participate in the development of eGov services and align them with user profiles capable of being mainstreamed across the North Sea's regional innovation system.

The unique nature of this academic network lies with understanding that triple helix models are not just about offering theoretically informed research and technical development opportunities, but a methodology equally capable of accounting for the social capital of the knowledge base available for communities to learn about how cities can be(come) smart. This is because for SCRAN, research defined solely in terms of scientific and technical development is not the network's common denominator. The reason for this is because the particular terms of reference it works to lie elsewhere and with the academic contribution that the network makes to the intellectual capital of the SmartCities venture. In particular with the advantage this academic network manages to construct as the social capital (Halpern, 2005), environmental and cultural assets of the knowledge base that underpins this community (Deakin and Allwinkle, 2005) and which is built to support their learning (Riemer and Klein, 2008).

Set within such terms of reference, SCRAN's particular task is to search out the potential advantage that the intellectual capital of the knowledge base underlying this learning community is able to build as a platform that supports wealth creation. In that sense: a platform supporting the wealth creation of industry and which is in turn regulated by government in terms of the electronically enhanced services being developed for such purposes.

Configuring the SmartCities triple helix

Figure 9.2 sets out SCRAN's attempt to meet the methodological challenges that such a process of knowledge production raises and offers an initial representation of the triple helix that SmartCities advances for such purposes. In semantic terms the three institutional dimensions of universities, industry and government are represented as the intellectual capital, wealth creation and regulation of eGov service developments and as that process of

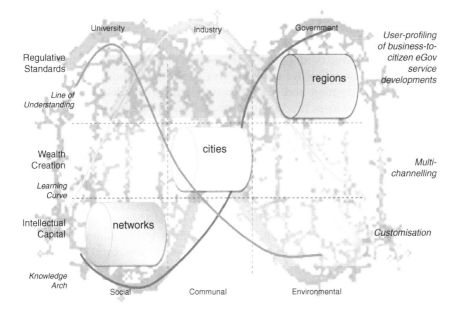

University   Industry   Government   *User-profiling*
                                                    of business-to-
Regulative   *citizen eGov*
Standards   regions   *service*
                                                    developments

Line of
Understanding

Wealth   cities   *Multi-*
Creation   *channelling*

Learning
Curve

Intellectual   *Customisation*
Capital   networks

Knowledge
Arch   Social   Communal   Environmental

Figure 9.2 The triple helix of SmartCities 1

Notes

1. Unlike most triple helix studies that rest upon the knowledge economy of information society, this representation is not set in the fundamental, or strategic, research
   ('blue-sky' or 'applied') domain of either a predominantly science and technology, or
   entrepreneurial-led university, but the intellectual capital which is generated from their
   'third mission' outreach into social networks as the knowledge base of wealth creation.
2. The intellectual capital embedded in these social networks offers the technological
   basis to begin understanding how industry organises communities so it becomes
   possible for them to learn about how they can manage the knowledge invested in
   this process of wealth creation.
3. The intellectual capital of this wealth creation in turn calls for government
   involvement in setting the standards for eGov service developments to regulate
   this process.
4. This wealth of intellect in turn offers the depth of academic understanding, social
   learning and communal knowledge needed for the development of eGov services to
   regulate all of this as part of a regional innovation system.
5. This is why the focus of this triple helix is on the knowledge base of those learning
   platforms supporting cities, rather than on understanding the standards of regulation governing regional innovation systems. That knowledge and learning which
   takes place here by way of cities and through the industry of their wealth creation.
   The industry of their wealth creation which is seen to be smart for the reason it is
   these communities that offer the means (intellectual capital of socially embedded
   networks and industry of wealth creation) to understand the nature of the developments in question. That is, as the network's customisation, cities' multi-channelling
   and their user-profiling of those businesses-to-citizen applications that make up this
   regional innovation system.
6. In many respects this (re)modelling of the triple helix as a process of knowledge
   production, generated by way of networked communities and through the wealth
   created from their industry, but on a set of standards integral to the governance of
   their regional innovation system, does what Jensen and Bjorn (2004), Jauhiainen
   and Suorsa (2008) and Smith (2007) ask of it.

knowledge production which is part of the North Sea's regional innovation system. Set out as an actor-network matrix of such institutional relations, it is universities, industry and government which make up the columns of the matrix, and their respective contributions to the generation of intellectual capital, wealth creation and regulative standards of the developments that make up the knowledge production of the left-hand row.

While the specific guidance available from Etzkowitz and Leydesdorff (2002, 2003) is particularly limited here, Figure 9.2 does take a lead from Leydesdorff and Cooke (2006) and Etzkowitz's (2008) configurations of the triple helix and 'geometries' of the model's reflexive qualities. Figure 9.2 configures these qualities as a three-by-three matrix. That is, as the building blocks of a social network having a knowledge base which generates the intellectual capital needed for the wealth created by SCRAN to be smart in making it possible for the ICTs supporting the industry of this community to be applied as the practical means by which both the environmental and cultural assets of such a virtual learning organisation can begin to meet the governance requirement. In particular, the requirement for knowledge products of this kind, i.e. those quintessentially civic in nature and developed on behalf of the public, to be regulated by standards set by government.

The step-wise logic

This first institutional step into a formal representation of SCRAN's triple helix is then given content by means of the analytical spaces the matrix opens up for the SmartCities venture and the opportunity this in turn offers the three-way partnership to cut across this as part of the North Sea's regional innovation system.

This networking of SmartCities as a regional innovation system in turn relates the universities engaged in the generation of intellectual capital, industry involved in the creation of wealth and government regulating the standards of the service development (i.e. in this instance, the generation and wealth from the development of eGov services as knowledge products) back to those actors not only associating with one another as a loosely cast network, but as a tightly knit and far-flung community, openly relating to itself and others as a virtual learning organisation. Step two of the representation captures this in terms of the wealth created by such a process of knowledge production; what social learning this community contributes is represented in the right-hand column of the matrix. This is shown in terms of the advantage that smart cities build as a platform of wealth creation and the environments this in turn cultivates to underpin the standards of eGov service developments set by the public to regulate this process of knowledge production. All of this is captured as step three and is represented in the far right-hand column. This is shown in terms of what the wealth of knowledge produced contributes to the development of eGov services as part of a regional innovation system. That is to say, by way of and through the associated capital and

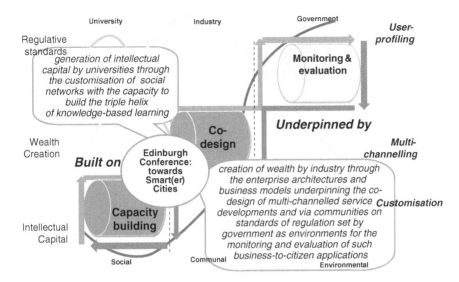

Figure 9.3 The triple helix of SmartCities 2

knowledge base of the learning community set up to promote the customisation of eGov services, the wealth created from their co-design and regulation of this process by the public in accordance with the user profiles which serve to both monitor and evaluate this development of the North Sea's regional innovation system.

Inverting the normal representation

Revealing how the triple helix of the SmartCities venture can be assembled, not only as a CoP but as a virtual learning organisation, equally constructive in what its intellectual capital contributes to the creative wealth of knowledge production is, however, not so simple. This is because proving that it needs to be grounded in policies which are socially inclusive, equitable and justly participative requires the academic network to accept the value of first 'inverting' the normal representation of the model's institutional relations. For without 'turning things upside down', it is not possible for the intellectual capital of universities to be deployed in this innovation system as so-called 'third mission' exercises, not only reaching out towards, but bottoming out the social networks that make up the learning communities that industry organises as the knowledge base of smart cities and applies as a set of standards by which the development of electronically enhanced services are regulated across the region.

Figure 9.3 offers SCRAN's attempt to do just this by presenting a second-order configuration of the triple helix for SmartCities. For this configuration shows the university as being responsible for building the capacity of the enterprise architecture and business models to act as a platform for cities to be smart in co-designing the development of eGov services with customised, multi-channelled access, targeting specific user profiles as components of the North Sea's regional innovation system.

Represented in this way, it is possible to be specific about the institutional duties and responsibilities of SCRAN's triple helix. For as Figure 9.3 shows, while the work packaged together under the titles of customisation, co-design and user profiling provides the backdrop to SCRAN, it is not proposed that SmartCities should cover all of them as components of the North Sea's regional innovation system. Rather, it suggests SCRAN should use the triple helix model as a means to cut across them, concentrating the efforts of the network's knowledge base on learning about building the capacity needed for cities to be smart in supporting the co-design, monitoring and evaluation requirements of eGov service development programmes. This is because the capacity building that underlies all of the aforesaid, and the generation of intellectual capital which their social networks support, make up the knowledge base of the community in question, the pedagogy this learning organisation constructs and the post-transactional logic that the industry's enterprise architecture and related business models in turn create. That architecture and model which is of particular significance in the sense that it allows such an enterprise to reach beyond itself and up into an environment where the knowledge it cultivates is not for profit, but quintessentially civic, lying in the public domain and the subject of governance. In particular, the subject of a governance that aligns itself with the standards of social inclusion, equity and participation which serve to regulate such developments on behalf of the public.

The social significance of such an articulation of the triple helix can perhaps best be captured in the opportunity it provides for the intellectual capital of universities and the wealth creation of industry not to foreclose on such possibilities and instead openly channel communications within environments that not only transcend the corporate sector, but achieve this by cultivating a regional innovation system which serves the public interest.

Organised in this way, it is possible to see the geometry of the 'knowledge-arch' and 'learning curve' that underlies SCRAN's take on the triple helix and which it lays down for understanding this venture. What this also serves to illustrate is the academic network's particular take on such an institutionalisation of the model. In particular the fact that the triple helix builds off the social network of a given knowledge base and allows universities to take the lead in generating the wealth of industry underpinning this by way of their enterprise architecture and through the business model supporting the SmartCities venture – those enterprise architectures and business models considered to be particularly important to SCRAN because they offer not only a platform for the associated capital of the communities to learn about what the customisation, co-design and multi-channelling of services means, but also a way to monitor and evaluate the implementation of such eGov developments as part of a regional innovation system

As already suggested, in order for universities to be part of something more than an informal social network and a more tightly knit community, this means that the academic content of the capital associated with the virtues of any such learning organisation demands a pedagogy. In particular, a pedagogy capable of constructing a knowledge of what is needed for the regulation of the wealth created to be advantageous in meeting the requirements of the SmartCities venture. The content of this learning community in turn provides the platform for what might best be termed the critical 'building-blocks' of smart cities and the eGov service development this offers. As critical components of SmartCities they all need to be linked together. It is the networking of the social capital underlying this process of knowledge production that might best be defined as the SmartCities learning community and whose regional innovation system the following section shall report on.

The network

The key factors distinguishing SCRAN from other networks are these:

* Here university engagement is not a top-down exercise in the generation of intellectual capital, or the creation of wealth, but the social capital of knowledge production by communities learning about how the development of eGov services can regulate this process.
* As such their involvement can be said to be bottomed out on the networking of social capital grounded within the third mission logic of participation.
* Undercutting previous representations of the helix, the object of the exercise might be said to be that of using the networking possibilities that social capital offers the SmartCities venture to make their learning communities the 'generators of intellectual wealth'.
* Networked as the associated capital of Web 2.0 technologies, the aim is for their learning communities to work smarter not harder. Smarter not harder in generating intellectual capital and creating wealth not gauged in terms of their economic worth, but as that which it is possible for cities to deploy as the means to govern their regional innovation systems. That intellectual capital and wealth creation which is of particular value for the reason this generates the means by which it becomes possible for the knowledge produced by learning communities to be codified. Not left as tacit knowledge to be exploited as the regular events of routine practices, but made explicit as codes with sufficient critical insight to be exceptional in the sense they build the capacity needed to override economic interests and grant civil society the power that is required for communities to learn about how the architectures they model can begin constructing

such business-to-citizen applications within environments able to regulate the standards by which such developments are governed. By which such developments are governed, not just on behalf of the public, but for the reason that like everything else in such developments, they too are integral parts of their regional innovation system.

The shared enterprise and joint venture

While this vision of SCRAN may be the enterprise shared across the partnership, the collaboration underpinning such a joint venture needs to be constructed through consensus building. As communications lie at the centre of the collaboration and it is web services that support the consensus building needed for cities to work smarter, not harder, it is the construction of these platforms that is the critical requirement of this shared enterprise and joint venture. This is because such platforms are pivotal in making it possible for the members of networked communities to learn about what works in the development of eGov services and relay a knowledge of both the wealth creation and governance opportunities underlying this onto others as part of the 'democratic body' supporting such action. Otherwise, the democratic qualities attending the social body of these spaces tend to be overlooked by those prioritising the community and the environmental components that such developments cultivate. The following argues that the 'demos' of such ventures is something that should not be ignored because their body of knowledge includes practices which are not only representative, but directly participative. The three dimensions of the venture are summarised in Figure 9.4.

Figure 9.4 illustrates that networking has been used as the means by which to collaborate and build consensus (primarily by SCRAN) over the service developments (by other work packages within the project). This in turn has built the capacity to service the co-design and monitoring and evaluation needs of their respective knowledge management requirements.

It is possible to explore these service developments further. First, the heading of capacity building focuses attention on the technical issues and, in particular, those supporting the partnership's enterprise architecture and business modelling. From here attention turns to the co-design of the customised and multi-channelled distribution of the eGov services. Finally, in terms of the monitoring and evaluation, questions of particular relevance surfacing here relate to the degree to which the eGov services developments raise standards of provision by:

- improving the quality of service provision through an enhanced customer experience;
- better matching customer needs with the required services;
- widening access to the services (via multi-channelled distribution) and being socially inclusive in deciding who they are directed at;

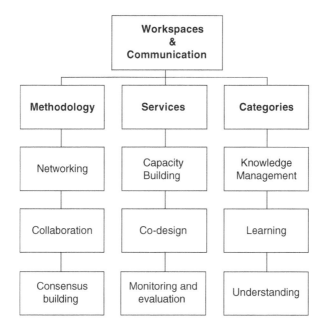

Figure 9.4 The three dimensions of the workspaces and communication

- being more efficient (cost-effective), just, participative and, thereby, democratically accountable in delivering the services relative to other modes of provision.

To meet these standards four web-service developments are required:

- information provision;
- collaboration and consensus building;
- knowledge management (KM);
- learning and (shared) understanding.

The SCRAN wiki

While a variety of collaborative platforms are available, the consensus was that SCRAN needed to put information systems in place that are simple, lightweight and robust. Simple, lightweight and robust in their use of available technology, but sufficiently agile to meet the growing requirements of the venture's learning community. The first need is met through a content management system (CMS), provided by Drupal.[2]

The communications needs and tools used by SmartCities are summarised in Figure 9.5. Fortunately it is now possible to offer the smart city partnership

Requirement	Tool used
Information:	
Public web-site	Drupal
News and events	Epractice.eu
Collaboration & consensus building	
Transient messages	Email, mailing list
Record of meetings	Drupal (login required)
Activity planning	Private Mediawiki pages
Discussions	Email
Formal decisions	(Face to face at project meetings)
Knowledge management	
Glossary	Mediawiki pages, cross linked to work package and expertise
Digital library	Documents stored on Mediawiki or Drupal
Search engine	Public pages: Drupal and Google search
	Private pages: Mediawiki search, use of categories, automatic cross-linking
Learning and understanding	
Good practice	Private Mediawiki pages
Case studies	Private Mediawiki pages
Use cases	Private Mediawiki pages + supporting documents

Figure 9.5 Choice of technology to support the needs of SmartCities

all of these web services by way of a wiki. In this regard Mediawiki[3] has been selected by SCRAN for its flexibility in supporting the network's knowledge management and learning requirements. This software allows users to freely create and edit web page content using any web browser. Wikis also support hyperlinks and have simple text syntax for creating new pages and cross-links between internal documents. Wikis are unusual communication systems in the sense that their organisation allows contributions to be edited. Allowing users to create and edit any page in a website is innovative in the way it allows civil society to democratise the use of the internet.

Evolution of the knowledge base

The wiki was initially designed to support communication between the SCRAN and the other SmartCities partners. This challenge has been met by developing a glossary to help with the terminology of eGov service developments, and a questioning framework supported by a set of research briefs (see Figure 9.6).

This architecture is also reflected in the relatively simple initial home page illustrated in Figure 9.7. From the start, the need to be disciplined in the use

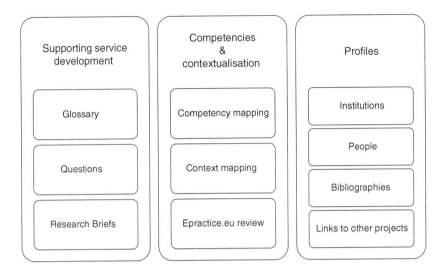

Figure 9.6 Initial architecture of the SmartCities wiki

Figure 9.7 Initial layout of the project wiki

of categories and structured templates has been recognised. The first task carried out was a competence mapping of the SCRAN partners, which provides a direct link between the partners and the work that is done to contextualise the developments. These are adopted as the collective competencies of SCRAN, the knowledge and learning this organisation contributes to the SmartCities partnership.

Secondly, a context mapping exercise was performed to gain an insight into the issues underlying the evolution of eGov services for each municipal partner, to identify how the eGov service developments are impacting on governments. This has been done to capture the institutionalisation of the eGov policy agenda by universities and industry, to highlight the drivers for change, agencies responsible for bringing about improvements in public service provision and the specific obligations this places on the cities to deliver on such expectations. Thirdly, a review of alternative sources of knowledge management and learning was carried out by examining case studies stored on the database maintained by epractice.eu. This is a portal for practitioners working in e-government and promoted by the EU as a benchmark of current knowledge and understanding.[4]

As the project has progressed, the use of the wiki has evolved. Merely looking at the home page at the start of 2010 (Figure 9.8) shows how the content of the wiki has become richer as the project has progressed and the practitioners have become engaged in the process.

Case study review

Epractice.eu is a major resource and dissemination tool for European projects. As an exercise, SCRAN explored and assessed the data it held on cases to highlight good practice relevant to the development of eGov services in the North Sea region. This found 30 potentially relevant cases that were reviewed for quality and relevance to the SmartCities venture, as a result of which the data set was pared back to 16. Out of the 16, only one referenced the triple helix model of eGov service developments.[5]

Figure 9.9 shows the frequency of occurrence of keywords in the investigated cases. It is evident that the common denominators with the SmartCities venture are their focus on policy, e-Gov services, multi-channel access, user-centric services and interoperable infrastructure technologies. Perhaps more significant is the lack of any reference to the enterprise architecture and business model of any such open system, or any notion of what it means to customise eGov services.[6] For while the case studies go some way to highlight the need for core infrastructure developments (i.e. secure access and safety of personal data), it is evident they offer little by way of critical insight into the underlying methodology of the inclusive policy agenda they flag up for the multi-channelling and user-profiling services drawn attention to.

This lack of critical insight is significant insofar as it raises particular questions about civil society's current trust in the ability of electronically enhanced

Figure 9.8 The wiki home page after a year

services to be secure enough for them do anything but store data and transfer information. In that sense, for them to be used as anything but the types of service transactions that SmartCities propose to customise and offer multi-channel access to as eGov service developments. Those types of transactions that are particularly important for the reason that such a customisation in turn provides a platform for the type of consultations and deliberations that SmartCities seek to develop as a means for them to learn about the services that communities (citizen and business) expect from their providers.

To do this it is evident that attention needs to be thrown back onto the research briefs and the use of these to support the application of such an enterprise architecture and business model. In particular, an application that is capable of underpinning such customisation of eGov service developments as transnational pilots. Transnational pilots that not only support the custom-isation of eGov service developments, but do this by way of the consultations

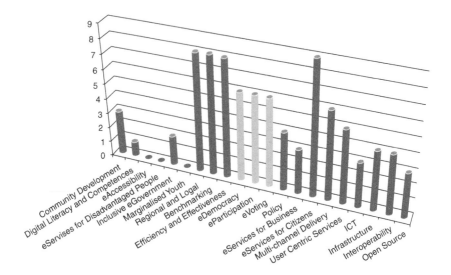

Figure 9.9 Frequency of self-labelled categories used by epractice.eu cases

and deliberations through which the multi-channel access and user profiling of such business-to-citizen applications in turn emerge as key components of their regional innovation system.

Commissioning of work

SCRAN also oversees a fund which is available to meet the cost of commissioning research and consultancy work from enterprises within the business sector which support the development of the eGov service agenda. This fund is designed to take the triple helix model full circle and commission work from university and industry sectors capable of consulting on the use of ICTs as a means of networking social capital. The work commissioned aligns with the enterprise architecture and business modelling needs of the capacity building, co-design, evaluation and monitoring requirements underlying the SmartCities venture. This in turn allows the work commissioned to:

- operationalise a triple helix model of eGov service developments;
- draw upon this knowledge base as a platform to build upon such eGov service developments;
- use the collective memory of this learning community as a means to instil a sense of identity, and use this as a way of embodying the wisdom attached to three-way partnerships of this kind;
- deploy this wisdom as part of a step change in eGov service developments and as a means to mainstream them as part of the North Sea's regional innovation system.

Conclusions

This chapter has focused on SCRAN's model of the triple helix for the SmartCities venture and drawn attention to the communication needs and organisational and technical requirements of the network. In particular, those needs that require to be met for the (inter) regional academic network to be smart in transferring knowledge about eGov development programmes between cities.

In meeting this aim, the chapter has reported on the three-way partnership working in the SmartCities venture, the methodology underlying the development of SCRAN as a network, the web services supporting the organisation's efforts, the commissioning of work from the business sector and the governance of this particular development process. It has gone on to configure SCRAN's triple helix and set out the 'step-wise' logic of the organisation's knowledge base and learning platform. From here attention has turned towards the networking of the triple helix by way of SCRAN's wiki and through the learning community supporting the SmartCities venture.

These developments are seen to be significant because under this representation of the triple helix, university engagement is no longer a top-down exercise in either the generation of intellectual capital or the creation of wealth by industry, but a bottom-up exercise in building the social capital regulating the development of eGov services. Such involvement can be said to be bottomed out on the networking of social capital and out with the normal domain of the triple helix. This in turn tends to undercut previous representations of the triple helix, so much so that here the object of the exercise is no longer the production of a knowledge economy, but the social capital of learning communities.

Here the social capital is embedded in the Web 2.0 technologies and their learning community's ambition to use them as a means to work smarter not harder. This in turn makes it possible for the knowledge produced to be codified. That is to say, made explicit as codes with sufficient critical insight to override economic interests and grant civil society sufficient power to govern over the electronically enhanced service developments that the regional innovation system is part of. Making all of this explicit also allows SCRAN to improve the quality of service provision by way of such business-to-citizen applications and through a process of customisation available over the web via multi-channel access and in line with the user profiles governing developments of this type.

All this does nothing less than operationalise SCRAN's triple helix model by allowing the network to use the capacities of the communities that it creates as the platform by which to learn about the development of eGov services. This also serves to standardise the transformation of eGovernment as a triple helix of Web 2.0-based knowledge infrastructures, architectures and enterprises. Knowledge infrastructures, architectures and enterprises that in turn allow organisations to learn about the critical role that business-to-citizen applications play in making it possible for cities to become smart in reaching beyond the transactional logic of service provision and by grasping the potential that the regional innovation systems offer the public to democratise this transformation.

This possibility is something that has all too often been missed by those reporting on the democratic transformation of eGovernment services currently underway, and yet it is critical in understanding whether the generation of intellectual capital by universities allows cities to be smart in creating wealth and if industry's deployment of such business-to-citizen applications, as the democratic means to govern regional innovation systems, are developments capable of regulating such exchanges. This offers an image of the triple helix that SCRAN proposes to develop. For it not only goes some way to capture this organisation's particular take on the triple helix, but also serves as a means for this CoP to draw upon as the scientific and technological basis of the SmartCities venture.

Notes

1 www.northsearegion.eu/ivb/projects/details/&tid=84
2 www.drupal.org
3 www.mediawiki.org
4 A secondary motivation is that epractice.eu also provides the primary route for Europe-wide dissemination of project activities.
5 By the Swedish innovation body INNOVA, www.vinnova.se
6 Again only one case provides evidence of a service-orientated enterprise architecture for eGov developments and there are no references to any underlying business model.

References

Amin, A. and Cohendet (2004) *Architectures of Knowledge: Firms, Capabilities, and Communities*, Oxford, Oxford University Press.

Amin, A. and Roberts, J. (2008) Knowing in Action: Beyond Communities of Practice, *Research Policy*, 37 (2): 353–369.

Asheim, B., Coenen, L., Moodysson, J. and Vang, J. (2007) Constructing Knowledge-based Regional Advantage: Implications for Regional Innovation Policy, *International Journal of Entrepreneurship and Innovation Management*, 7 (2–5): 140–155.

Asheim, B., Cooke, P. and Martin, R. (2006) Clusters and Regional Development: Critical Reflections and Explorations, *Economic Geography*, 84 (10): 109–112.

Caragliu, A., Del Bo, C. and Nijkamp, P. (2009) Smart Cities in Europe, *Serie Research Memoranda 0048*, VU University Amsterdam, Faculty of Economics, Business Administration and Econometrics. Retrieved from http://ideas.repec.org/p/dgr/vuarem/2009-48.html.

Cooke, P. (2007) Regional Innovation Systems, Asymmetric Knowledge and the Legacies of Learning, in R. Rutten, F. Boekema and G. Hospers, eds, *The Learning Region: Foundations, State of the Art, Future*, Cheltenham, Edward Elgar.

Deakin, M. (2009) The IntelCities Community of Practice: The eGov Services Model for Socially-inclusive and Participatory Urban Regeneration Programmes, in

Reddick, C. ed., *Research Strategies for eGovernment Service Adoption*, Hershey, Idea Group Publishing.

Deakin, M. (2010) SCRAN's Development of a Trans-national Comparator for the Standardisation of eGovernment Services, in Reddick, C., ed. *Comparative E-Government: An Examination of E-Government Across Countries*, Berlin, Springer Press.

Deakin, M. and Allwinkle, S. (2005) The IntelCities eLearning Platform, Knowledge Management System and Digital Library for Semantically Interoperable e-Governance Services, in Cunningham, P., ed., *Innovation and the Knowledge Economy: Issues, Applications and Case Studies*, Washington, DC, ISO Press.

Deakin, M., Allwinkle, S. and Campbell, F. (2006) The IntelCities e-Learning Platform, Knowledge Management System and Digital Library for Semantically Rich e-Governance Services, *International Journal of Technology, Knowledge and Society*, 2 (8): 31–38.

Deakin, M., Allwinkle, S. and Campbell, F. (2008) The IntelCities Community of Practice: The eGov Services for Socially-inclusive and Participatory Urban Regeneration Programmes, in Cunningham, P., ed., *Innovation and the Knowledge Economy: Issues, Applications and Case Studies*, Washington, DC, ISO Press.

Etzkowitz, H. (2008) *The Triple Helix: University-industry-government Innovation in Action*, Oxford, Routledge.

Etzkowitz, H. and Leydesdorff, L., eds (2002) *Universities and the Global Knowledge Economy NIP: A Triple Helix of University-industry-government Relations*, Continuum International Publishing Group Ltd.

Etzkowitz, H. and Leydesdorff, L. (2003) Can 'the Public' Be Considered as a Fourth Helix in University-Industry-Government Relations? Report of the Fourth Triple Helix Conference, *Science & Public Policy*, 30 (1): 55–61.

Gibbons, M., Limoges, C., Nowotny, H. and Schwartzman, S. (1994) *The New Production of Knowledge: The Dynamics of Science and Research in Contemporary Societies*, London, Sage Publications.

Halpern, D. (2005) *Social Capital*, Bristol, Polity Press.

Jauhiainen, J. and Suorsa, K. (2008) Triple Helix in the Periphery: The Case of Multipolis in Northern Finland, *Cambridge Journal of Regions, Economy and Society*, 1 (2): 285–301.

Jensen, J. and Bjorn, T. (2004) Narrating the Triple Helix Concept in 'Weak' Regions: Lessons from Sweden, *International Journal of Technology Management*, 27 (5): 513–530.

Leydesdorff, L. and Cooke, P. (2006) Regional Development in the Knowledge-Based Economy: The Construction of Advantage, *Journal of Technology Transfer*, 1–15.

Leydesdorff, L. and Henry Etzkowitz, H. (2001) The Dynamics of Innovation: From National Systems and 'Mode 2' to a Triple Helix of University–industry–government Relations, *Research Policy*, 29: 109–123.

Riemer, K. and Klein, S. (2008) Is the V-form the Next Generation Organisation? An Analysis of Challenges, Pitfalls and Remedies of ICT-enabled Virtual Organisations Based on Social Capital Theory, *Journal of Information Technology*, 23: 147–162.

Smith, H. (2007) Universities, Innovation, and Territorial Development: A Review of the Evidence, *Environment and Planning C*, 23 (1): 89–114.

Wenger, E. (1998) *Communities of Practice: Learning, Meaning, and Identity*, Cambridge, Cambridge University Press.

Part III
Analysing the transition

10 Smart cities in Europe

*Andrea Caragliu, Chiara Del Bo and
Peter Nijkamp*

Introduction

What is the source of urban growth and of sustainable urban development? This question has received continuous attention from researchers and policy makers for many decades. Cities all over the world are in a state of flux and exhibit complex dynamics. As cities grow, planners devise:

> complex systems to deal with food supplies on an international scale, water supplies over long distances and local waste disposal, urban traffic management systems and so on; […] and the quality of all such urban inputs defines the quality of life of urban dwellers. (Science Museum, 2004)

Notwithstanding the enormous formidable challenges and disadvantages associated with urban agglomerations, the world's population has been steadily concentrating in cities. Figure 10.1 shows the percentage of EU citizens living in cities (the population living in areas classified as urban according to the country-specific criteria selected by the UN); a massive rise in this percentage took place, from slightly more than 50 per cent in 1950 to more than 75 per cent of EU population being located in urban areas in the year 2010, and a forecast of about 85 per cent within the next 40 years.

In addition, we also witness a substantial increase in the average size of urban areas. This has been made possible by a simultaneous upward shift in the urban technological frontier, so that a city can accommodate more inhabitants. Problems associated with urban agglomerations have usually been solved by means of creativity, human capital, cooperation (sometimes bargaining) among relevant stakeholders and bright scientific ideas: in a nutshell, 'smart' solutions. The label 'smart city' should therefore point to clever solutions allowing modern cities to thrive, through quantitative and qualitative improvement in productivity. However, when googling 'smart city definition',[1] we discover that among the very first results we can name a communications provider, a US radio, an Edinburgh hostel, an initiative of the Amsterdam Innovation Engine and so on; but no sign of a proper definition.

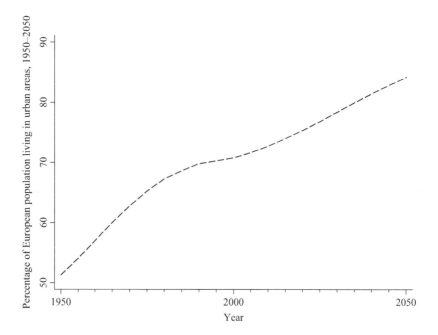

Figure 10.1 Percentage of EU population living in urban areas, 1950–2050 (forecast)
Source: UN (2009)

In this chapter we search for a clear and focused definition of the label 'smart city'. We next provide qualitative evidence on the correlations between the dimensions of our definition of smart cities and a measure of wealth, i.e. per capita GDP in purchasing power parity (henceforth, PPP).[2] We will start with a brief literature review.

Literature review

The concept of the 'smart city' has become quite fashionable in the policy arena in recent years. Its main focus seems to be on the role of ICT infrastructure, although much research has also been carried out on the role of human capital, social and relational capital and environmental interest as important drivers of urban growth.

The European Union (EU), in particular, has devoted constant efforts to devising a strategy for achieving urban growth in a 'smart' sense for its metropolitan areas. Not only the EU, but also other international institutions and think tanks believe in a wired, ICT-driven form of development. The Intelligent Community Forum, for instance, produces research on the local effects of the ICT revolution, which is now available worldwide. The OECD and EUROSTAT *Oslo Manual* (2005) stresses instead the role of innovation in ICT sectors and provides a toolkit to identify consistent indicators, thus shaping a sound framework of analysis for researchers on urban innovation. At a meso-regional level, we observe renewed attention for the role of soft communication infrastructure in determining economic performance.[3]

The availability and quality of the ICT infrastructure is not the only definition of a smart or intelligent city. Without reference to the 'smartness' concept, the relation between ICT infrastructure and economic performance has been an object of academic interest since the beginning of the digital era (e.g. Roller and Waverman, 2001). Other definitions stress the role of human capital and education in urban development. Berry and Glaeser (2005) and Glaeser and Berry (2006) show, for example, that the most rapid urban growth rates have been achieved in cities where a high share of educated labour force is available. In particular, Berry and Glaeser (2005) model the relation between human capital and urban development by assuming that innovation is driven by entrepreneurs who innovate in industries and products which require an increasingly more skilled labour force.

As not all cities are equally successful in investing in human capital, the data shows that an educated labour force – or, in Florida's jargon, the 'creative class' – is spatially clustering over time. This recognised tendency of cities to diverge in terms of human capital has attracted the attention of researchers and policy makers. It turns out that some cities, which were in the past better endowed with a skilled labour force, have managed to attract more skilled labour, whereas competing cities failed to do so. Policy makers, and in particular European ones, are most likely to attach a consistent weight to spatial homogeneity; in these circumstances the progressive clustering of urban human capital is then a major concern.

An interesting contribution (Fu, 2007) relates the smartness concept to the generation of localised knowledge spillovers (LKS). In this paper, human capital externalities originate from face-to-face contacts between peers in an urban environment. This paper follows the traditional literature on LKS, which includes Rauch (1993). Recent and valuable critical reviews of the concept of LKS can be found in Breschi and Lissoni (2001) and in Capello (2009).

The label 'smart city' is still, in our opinion, quite a fuzzy concept. Hollands (2008) stresses this point while also providing several examples of self-defined smart cities. In this chapter, we move forward by adding a critical review of the literature on smart urban growth from an economist's perspective and an exploratory empirical analysis. With this aim, we summarise the characteristics of a smart city that tend to be common to many of the previous findings as follows:

1. The 'utilization of networked infrastructure to improve economic and political efficiency and enable social, cultural and urban development',[4] where the term 'infrastructure' indicates business services, housing, leisure and lifestyle services, and ICTs (mobile and fixed phones, computer

networks, e-commerce, internet services). This description brings to the forefront the idea of a wired city as the main development model and of connectivity as the source of growth.

2. An 'underlying emphasis on business-led urban development'. According to several critiques of the concept of the smart city, this idea of neo-liberal urban spaces, where business-friendly cities would aim to attract new businesses, would be misleading. However, although caveats on the potential risks associated with putting an excessive weight on economic values as the sole driver of urban development may be worth noting, the data actually shows that business-oriented cities are indeed among those with a satisfactory socio-economic performance.

3. A strong focus on the aim to achieve the social inclusion of various urban residents in public services (e.g. Southampton's smartcard; see Southampton City Council, 2006). This prompts researchers and policy makers to give attention to the crucial issue of equitable urban growth. In other words: to what extent do all social classes benefit from a techno-logical impulse to their urban fabric?

4. A stress on the crucial role of high-tech and creative industries in long-run urban growth. This factor, along with 'soft infrastructure' ('know-ledge networks, voluntary organizations, crime-free environments, after dark entertainment economy'), is the core of Richard Florida's research (Florida, 2002). The basic idea in this case is that 'creative occupations are growing and firms now orient themselves to attract "the creative". Employers now prod their hires onto greater bursts of inspiration. The urban lesson of Florida's book is that cities that want to succeed must aim at attracting the creative types who are, Florida argues, the wave of the future' (Glaeser, 2005). The role of creative cultures in cities is also critically summarised in Nijkamp (2010), where creative capital co-determines, fosters and reinforces trends of skilled migration. While the presence of a creative and skilled workforce does not guarantee urban performance, in a knowledge-intensive and increasingly globalised economy these factors will increasingly deter-mine the success of cities.

5. Profound attention to the role of social and relational capital in urban development. A smart city will be a city whose community has learned to learn, adapt and innovate (Coe et al., 2001). People need to be able to use the technology in order to benefit from it: this refers to the absorptive capacity literature. This concept has been applied to different economic relations at different levels of spatial aggregation. The basic reference is Cohen and Levinthal (1990); Abreu et al. (2008) bridge the idea from a micro-, firm level to a more aggregated, meso-level; finally, Caragliu and Nijkamp (2012) test the role of regional absorptive capacity in inducing spatial knowledge spillovers.

    When social and relational issues are not properly taken into account, social polarisation may arise as a result. This last issue is also linked to economic, spatial and cultural polarisation. It should be noted, however, that some research actually argues the contrary. Poelhekke (2006), for example, shows that the concentration of high skilled workers is conducive to urban growth, irrespective of the polarisation effects that this process may generate at a meso- (for example, regional) level. The debate on the possible class inequality effects of policies oriented towards creating smart cities is, however, still not resolved.

6. Finally, social and environmental sustainability as a major strategic component of smart cities. In a world where resources are scarce, and where cities are increasingly basing their development and wealth on tourism and natural resources, their exploitation must guarantee the safe and renewable use of natural heritage. This last point is linked to the third item, because the wise balance of growth-enhancing measures, on the one hand, and the protection of weak links, on the other, is a cornerstone for sustainable urban development.

Items 5 and 6 are for us the most interesting and promising ones, from both a research and a policy perspective; we believe, therefore, that they may represent the object of future research for urban economists. In the next sections we provide quantitative and analytical evidence on the role of the creative class and human capital in sustainable urban development, arguing that it is indeed the mix of these two dimensions that determines the very notion of a 'smart' city. The relational capital side of the story is not evaluated in the present chapter, but this will be the subject of further research in future studies.

Along with the previously mentioned critical points, additional critiques have been advanced to question the concept of a smart or intelligent city. Hollands (2008) provides a thorough treatment of the main arguments against the superficial use of this concept in the policy arena. His main points are the following:

• The focus of the concept of smart city may lead to an underestimation of the possible negative effects of the development of the new technological and networked infrastructures needed for a city to be smart (on this topic, see also Graham and Marvin, 1996).
• This bias in strategic interest may lead to ignoring alternative avenues of promising urban development.
• Among these possible development patterns, policy makers would better consider those that depend not only on a business-led model. As a globalised business model is based on capital mobility, following a business-oriented model may result in a losing long-term strategy: 'The "spatial fix" inevitably means that mobile capital can often "write its own deals" to come to town, only to move on when it receives a better deal elsewhere. This is no less true for the smart city than it was for the industrial, manufacturing city' (Hollands, 2008: 314).

From a US perspective, research on smart cities has also evaluated the relevance of smart urban development in fighting urban sprawl (Bronstein, 2009), used a cognitive approach in assessing the role of psychological and cognitive attitudes towards ICTs in reducing the extent of the digital divide (Partridge, 2004) and verified on the field (through a case study on a community project) whether concrete action can be taken against such a digital divide in poor urban areas (McAllister et al., 2005).

Our chapter will now provide some quantitative evidence on many of these points, supported by spatial statistics, maps and graphical evidence on each of the points that the literature on smart cities has put forward, in order to explore and identify statistical correlations with socio-economic urban performance.

An operational definition of the 'smart city'

A narrow definition of a much-used concept may help in understanding the scope of this chapter. Although several different definitions of 'smart city' have been given in the past, most of them focus on the role of communication infrastructure. However, this bias reflects the time period when the smart city label gained interest, namely the early 1990s, when ICTs first reached a wide audience in European countries. Hence, in our opinion, the stress on the internet as 'the' smart city identifier no longer suffices.

A recent and interesting project conducted by the Centre of Regional Science at the Vienna University of Technology identifies six main 'axes' (dimensions) along which a ranking of 70 European middle-size cities can be made. These axes are: a smart economy, smart mobility, a smart environment, smart people, smart living and, finally, smart governance. These six axes connect with traditional regional and neoclassical theories of urban growth and development. In particular, the axes are based – respectively – on theories of regional competitiveness, transport and ICT economics, natural resources, human and social capital, quality of life and participation of societies in cities. We believe this offers a solid background for our theoretical framework, and therefore we base our definition on these six axes.

We believe a city to be smart when investments in human and social capital and traditional (transport) and modern (ICT) communication infrastructure fuel sustainable economic growth and a high quality of life, with a wise management of natural resources, through participatory governance.

Quantitative and graphical evidence on European smart cities

In this section we will present graphical and quantitative evidence on the relative performance and rankings of European cities with respect to measures reflecting some of the definitions of a smart city given in the literature. The data source is the Urban Audit data set in its latest wave (2003–2006). The Urban Audit consists of a collection of comparable statistics and indicators

Figure 10.2 Cities in the 2003–2006 Urban Audit survey

for European cities; it contains data for over 250 indicators across the
following domains:

- demography
- social aspects
- economic aspects
- civic involvement
- training and education
- environment
- travel and transport
- information society
- culture and recreation.

Cities that were surveyed in the latest available wave are depicted in
Figure 10.2.

We now present a set of charts which show partial correlations between
urban growth determinants and our measure of economic output, which is
per capita GDP in purchasing power standards (PPS) in 2004 (the latest data

	Per capita GDP in PPS	Employment in the entertainment industry	Multimodal accessibility	Length of public transport network	eGovern-ment	Human capital
Per capita GDP in PPS	1					
Employment in the entertainment industry	0.215 (0.1258)	1				
Multimodal accessibility	0.7049 0	-0.0059 (0.9553)	1			
Length of public transport network	0.3104 (0.0043)	0.2874 (0.0302)	0.0919 (0.312)	1		
e-Government	0.1418 (0.1751)	-0.0254 (0.8385)	0.141 (0.1004)	-0.0339 (0.7417)	1	
Human capital	-0.1361 (0.265)	-0.0983 (0.3649)	0.0833 (0.3616)	-0.0741 (0.5946)	0.0665 (0.5733)	1

Figure 10.3 Partial correlations between the six indicators of smart cities

available in the Urban Audit data set). For the sake of readability, cities are indicated with their Urban Audit code, while a complete correspondence table is available as an Appendix to this chapter.

The set of all partial correlations among the variables we use to measure the 'smartness' of European cities can be found in Figure 10.3, with corresponding p-values in parentheses. It is evident that most of the variables that we deem as capable of both co-determining long-run urban performance and characterising a thorough definition of smart city, tend to be positively associated with our measure of urban wealth (we chose per capita GDP in PPS in 2004 in order to avoid the problem of size effects and to take into account price differentials across countries, which might be particularly different among EU15 and New Member State (NMS) cities). An interesting but puzzling result arises for the relationship between the level of education of people living in our sample and their average individual income; this issue will be further analysed later in this section. Throughout this section, on the map as well as in our charts, we indicate the code of the city associated with each observation. We believe this to be a useful tool of analysis for both researchers as well as policymakers, to identify intriguing spatial issues in the Urban Audit data set, the possible presence of country effects, and in general to allow the reader to identify the locational patterns of our smart city measures.

Figure 10.3 offers partial support for Richard Florida's arguments on the role of the 'creative class' in determining long-run urban performance. Positive

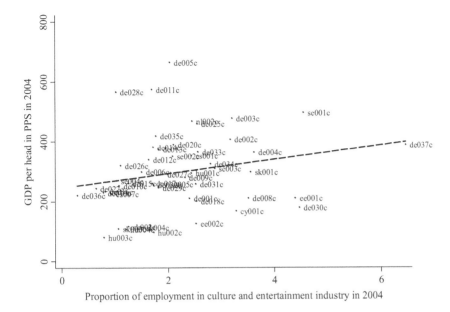

Figure 10.4 Creative class and wealth in 2004

correlations between the share of people employed in a 'creative' industry (Florida 2002, 2009), and in particular in the 'super-creative core', are found in US cities and states. In Florida (2002) the 'creative class' is defined as the merger of two Standard Occupational Classification System codes within the US labor force, namely:

- a super-creative core with those employed in science, engineering, education, computer programming and research, and with arts, design and media workers making a small subset. Those belonging to this group are considered to 'fully engage in the creative process' (Florida, 2002: 69);
- creative professionals with those employed in healthcare, business and finance, the legal sector and education.

Figure 10.4 measures these effects with the share of the labour force in European cities in the culture and entertainment industry; we find that the two measures show a positive and significant correlation (the correlation coefficient equals .2150 with a p-value of .1258).

In the urban economics literature, Florida's view has not been exempt from criticism (Glaeser, 2005). In the opinion of several economists, the argument that the creative professions would drive urban performance is flawed, and it would only be a proxy for the role of the 'hard' measurable stock of human capital (i.e. technical professions and total years of schooling) on

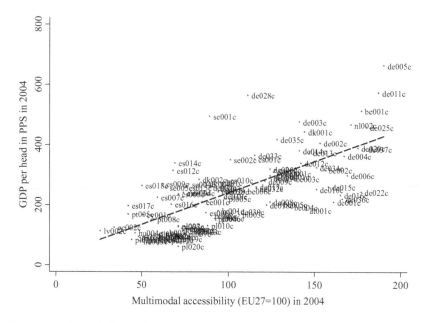

Figure 10.5 Accessibility and wealth in 2004

urban growth. Shapiro (2008) provides an excellent and convincing bridge between the two views. In his paper he proves with careful econometric estimations that human capital in cities contributes both directly to urban growth (measured by the growth of population, wages and two land rent measures) through productivity gains and indirectly through the increase in urban amenities, which in turn may foster the process of attraction of the creative class. Although the productivity effects are still the largest, according to Shapiro's estimates the amenities effects would account for as much as 20 to 30 per cent of total human capital effects on urban growth.

A second positive (and extremely significant) correlation appears to exist between multimodal accessibility and per capita GDP (Figure 10.5). The multimodal accessibility index is based on the assumption that the attraction of a destination increases with its size (in terms of population and GDP) and declines with distance, travel time and costs (which in turn lays its foundations in gravitational models of trade). Values of the index oscillate around 100, which is the average for the EU27. In this chart, the accessibility indicator, calculated as a weighted average of the ease with which a city can be reached with a combined set of available transportation modes (i.e. rail, road, sea or plane), also represents a measure for the market potential available to and from the city itself. Therefore, a better endowment of transportation might be conducive to wealth and growth, this last statement being in line with the New Economic Geography's theoretical expectations. For an excellent example

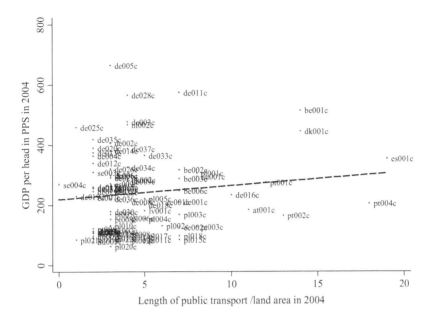

Figure 10.6 Public transport and wealth

of the role of the market potential in driving economic performance in the new economic geography literature, we refer to Redding and Sturm (2008), who in turn follow the rich tradition encompassing, among many, Davis and Weinstein (2003) and Hanson (2005).

Figure 10.6 shows instead the relationship between the availability of public transportation (normalised by the city area) and the level of wealth, measured as before with per capita GDP in PPS. The relationship is strongly positive; the city of Stockholm has been excluded from the original dataset as it behaves as an outlier, with an outstandingly high density of public transportation. With the inclusion of Stockholm the interpolation line would become even steeper. It is quite evident that an efficient net of public transportation is associated with high levels of wealth. Although the direction of causality in this relation may go both ways, it seems reasonable to think that a dense public transportation network may help to reverse the negative effects of urban density, thus at least partly releasing the pressure this exerts on the urban landscape and reducing the costs associated with congestion.

A slightly less significant and less steep association can be found between the level of GDP and a measure of e-government. The Urban Audit data set yields both the absolute number of government forms that can be downloaded from the website of the municipal authority, as well as the number of administrative forms which can be submitted electronically. As this last series has slightly more observations, and is, in our opinion, a better measure of the

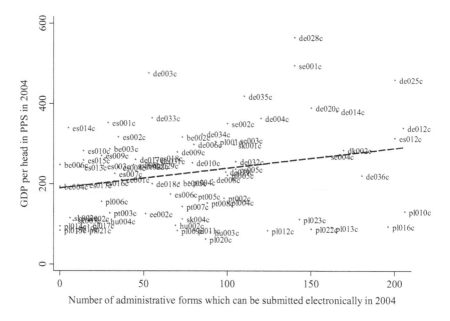

Figure 10.7 e-Government and wealth

real chance for citizens to interact with the urban public administration via the net, we represent this in Figure 10.7. The city of Krakow is in this case excluded as an outlier (in terms of number of forms that can be submitted online). The relationship does not change when the e-government measure is normalised by population or labour force (although this operation slightly alters the relative ranking of the cities in our sample).

Although cities with a high level of per capita GDP also tend to devote more attention to 'smart', e-government solutions, it is interesting to observe that some noticeable exceptions characterise this analysis. Some cities in peripheral countries (Krakow in Poland, Zaragoza in Spain, Ponto Delgada in Portugal) have also devised a wide set of forms that citizens can submit online, thus reducing travel and commuting costs, and costs associated with the management of multi-task public administration bodies.

Finally, Figure 10.8 shows the relationship between the stock of human capital and the level of urban wealth. According to neoclassical theories (Lucas, 1988; Arrow, 1962; Mankiw et al., 1992), human capital levels are good predictors of subsequent economic performance. In our sample this positive relationship has, nevertheless, more complex characteristics. The correlation coefficient between our measure of human capital, i.e. the share of the labour force qualified at ISCED levels 3 and 4, and the level of GDP is negative (although not significant at any statistical confidence level). As UNESCO's website explains:

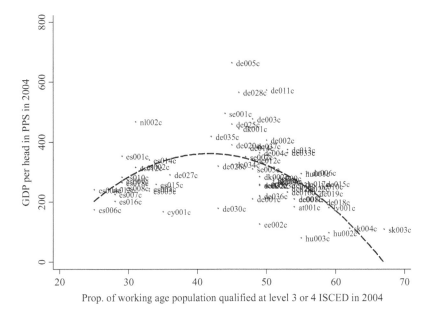

Figure 10.8 Human capital and wealth

The International Standard Classification of Education (ISCED) was designed by UNESCO in the early 1970s to serve 'as an instrument suitable for assembling, compiling and presenting statistics of education both within individual countries and internationally'. It was approved by the International Conference on Education (Geneva, 1975), and was subsequently endorsed by UNESCO's General Conference when it adopted the Revised Recommendation concerning the International Standardization of Educational Statistics at its twentieth session (Paris, 1978).

Does this imply that more education is associated with poorer economic conditions? If we look at Figure 10.8 it seems clear that the correct fit of this relationship is through a quadratic interpolation. After an appropriate (quadratic) term has been taken into account, the linear correlation between human capital and GDP is positive and significant at the 1 per cent level.[5]

The interpretation of this finding is, however, more difficult. By inspecting Figure 10.8 it is possible to make some observations on cities in the NMS of the EU at the right-hand side of the chart. As a legacy of the communist period, when levels of education were deliberately held high, labour forces in those countries may still own a large stock of human capital, although overall levels of individual wealth may not yet match those of the old member states. In this case, therefore, the depicted relationship may actually represent an off-saddle growth path portrait of the real human capital-urban growth equation. Indirect evidence to support this guess comes from splitting the sample into countries that in the 1980s were liberal or 'capitalist' in Europe and those that belonged to COMECON, and then fitting the data with a linear trend; the latter turns out to be positive and significant for the first of these two subsamples and negative and significant for the second.

A second key to interpret the puzzle may be obtained by reconnecting our study to Mayer (2007). She analyses the different ways in which cities and regions can set up a high-technology cluster even without the presence of a sound research-oriented university, whilst also criticising the opposite side of the story, namely the idea that academic research centres are a necessary and sufficient condition for achieving high-tech-oriented urban development. Therefore, cities in NMS may still fail to provide a sound connection between academic research institutes and the real economy, thus failing to attract the human capital-rich workers who raise productivity and wealth.

Conclusions and policy implications

In this chapter we have presented an overview of the concept of the 'smart city', with a critical review of the previous economics and planning approaches to this concept. We then presented a narrower definition of the concept of the smart city, and reviewed some quantitative and graphical evidence on the correlations of some of the main determinants of economic performance and the most important measure of urban success, namely per capita wealth.

Data from the 2004 Urban Audit data set show consistent evidence of a positive association between urban wealth and the presence of a vast number of creative professionals, a high score in a multimodal accessibility indicator, the quality of urban transportation networks, the diffusion of ICTs (most noticeably in the e-government industry) and, finally, the quality of human capital. These positive associations clearly define a policy agenda for smart cities, although clarity does not necessarily imply ease of implementation.

All variables shown to be positively associated with urban growth can be conceived of as stocks of capital; they are accumulated over time and are subject to decay processes. Hence, educating people is on average successful only when investment in education is carried out over a long period with a stable flow of resources; transportation networks must be constantly updated to keep up with other fast-growing cities, in order to keep attracting people and ideas; the fast pace of innovation in the ICT industry calls for a continuous and deep restructuring and rethinking of the communication infrastructure, to prevent European cities from losing ground to global competitors.

This continuous challenge, the 'endless frontier' to quote Vannevar Bush's words on scientific research (Bush, 1945), is the only way to ensure a sustainable path of development for cities, whilst at the same time guaranteeing that cities will maintain their crucial role as the cradle of ideas and freedom.

Appendix: Urban Audit codes and city names

Urban Audit code	City name
at001c	Wien
at002c	Graz
at003c	Linz
be001c	Bruxelles
be002c	Antwerpen
be003c	Gent
be004c	Charleroi
be005c	Liège
be006c	Brugge
bg001c	Sofia
bg002c	Plovdiv
bg003c	Varna
bg004c	Burgas
bg005c	Pleven
bg006c	Ruse
bg007c	Vidin
ch001c	Zürich
ch002c	Genève
ch004c	Bern
ch005c	Lausanne
cy001c	Lefkosia
cz001c	Praha
cz002c	Brno
cz003c	Ostrava
cz004c	Plzen
cz005c	Usti nad Labem
de001c	Berlin
de002c	Hamburg
de003c	München
de004c	Köln
de005c	Frankfurt am Main
de006c	Essen
de008c	Leipzig
de009c	Dresden
de010c	Dortmund
de011c	Düsseldorf
de012c	Bremen
de013c	Hannover
de014c	Nürnberg

Urban Audit code	City name
de015c	Bochum
de016c	Wuppertal
de017c	Bielefeld
de018c	Halle an der Saale
de019c	Magdeburg
de020c	Wiesbaden
de021c	Göttingen
de022c	Mülheim a.d.Ruhr
de023c	Moers
de025c	Darmstadt
de026c	Trier
de027c	Freiburg im Breisgau
de028c	Regensburg
de029c	Frankfurt (Oder)
de030c	Weimar
de031c	Schwerin
de032c	Erfurt
de033c	Augsburg
de034c	Bonn
de035c	Karlsruhe
de036c	Mönchengladbach
de037c	Mainz
dk001c	Copenhagen
dk002c	Aarhus
dk003c	Odense
dk004c	Aalborg
ee001c	Tallinn
ee002c	Tartu
es001c	Madrid
es002c	Barcelona
es003c	Valencia
es004c	Sevilla
es005c	Zaragoza
es006c	Málaga
es007c	Murcia
es008c	Las Palmas
es009c	Valladolid
es010c	Palma di Mallorca
es011c	Santiago de Compostela
es012c	Vitoria/Gasteiz

Urban Audit code	City name
es013c	Oviedo
es014c	Pamplona/Iruña
es015c	Santander
es016c	Toledo
es017c	Badajoz
es018c	Logroño
fi001c	Helsinki
fi002c	Tampere
fi003c	Turku
fi004c	Oulu
fr001c	Paris
fr003c	Lyon
fr004c	Toulouse
fr006c	Strasbourg
fr007c	Bordeaux
fr008c	Nantes
fr009c	Lille
fr010c	Montpellier
fr011c	Saint-Etienne
fr012c	Le Havre
fr013c	Rennes
fr014c	Amiens
fr015c	Rouen
fr016c	Nancy
fr017c	Metz
fr018c	Reims
fr019c	Orléans
fr020c	Dijon
fr021c	Poitiers
fr022c	Clermont-Ferrand
fr023c	Caen
fr024c	Limoges
fr025c	Besançon
fr026c	Grenoble
fr027c	Ajaccio
fr028c	Saint Denis
fr029c	Pointe-a-Pitre
fr030c	Fort-de-France
fr031c	Cayenne
fr032c	Toulon

Urban Audit code	City name
fr035c	Tours
fr202c	Aix-en-Provence
fr203c	Marseille
fr205c	Nice
fr207c	Lens – Liévin
gr001c	Athina
gr002c	Thessaloniki
gr003c	Patra
gr004c	Iraklio
gr005c	Larissa
gr006c	Volos
gr007c	Ioannina
gr008c	Kavala
gr009c	Kalamata
hu001c	Budapest
hu002c	Miskolc
hu003c	Nyiregyhaza
hu004c	Pecs
ie001c	Dublin
ie002c	Cork
ie003c	Limerick
ie004c	Galway
it001c	Roma
it002c	Milano
it003c	Napoli
it004c	Torino
it005c	Palermo
it006c	Genova
it007c	Firenze
it008c	Bari
it009c	Bologna
it010c	Catania
it011c	Venezia
it012c	Verona
it013c	Cremona
it014c	Trento
it015c	Trieste
it016c	Perugia
it017c	Ancona
it018c	l'Aquila

Urban Audit code	City name
it019c	Pescara
it020c	Campobasso
it021c	Caserta
it022c	Taranto
it023c	Potenza
it024c	Catanzaro
it025c	Reggio di Calabria
it026c	Sassari
it027c	Cagliari
lt001c	Vilnius
lt002c	Kaunas
lt003c	Panevezys
lu001c	Luxembourg
lv001c	Riga
lv002c	Liepaja
mt001c	Valletta
nl001c	s' Gravenhage
nl002c	Amsterdam
nl003c	Rotterdam
nl004c	Utrecht
nl005c	Eindhoven
nl006c	Tilburg
nl007c	Groningen
nl008c	Enschede
nl009c	Arnhem
nl010c	Heerlen
pl001c	Warszawa
pl002c	Lodz
pl003c	Krakow
pl004c	Wroclaw
pl005c	Poznan
pl006c	Gdansk
pl007c	Szczecin
pl008c	Bydgoszcz
pl009c	Lublin
pl010c	Katowice
pl011c	Bialystok
pl012c	Kielce
pl013c	Torun
pl014c	Olsztyn

Urban Audit code	City name
pl015c	Rzeszow
pl016c	Opole
pl017c	Gorzow Wielkopolski
pl018c	Zielona Gora
pl019c	Jelenia Gora
pl020c	Nowy Sacz
pl021c	Suwalki
pl022c	Konin
pl023c	Zory
pt001c	Lisboa
pt002c	Oporto
pt003c	Braga
pt004c	Funchal
pt005c	Coimbra
pt006c	Setubal
pt007c	Ponto Delgada
pt008c	Aveiro
ro001c	Bucuresti
ro002c	Cluj-Napoca
ro003c	Timisoara
ro004c	Craiova
ro005c	Braila
ro006c	Oradea
ro007c	Bacau
ro008c	Arad
ro009c	Sibiu
ro010c	Targu Mures
ro011c	Piatra Neamt
ro012c	Calarasi
ro013c	Giurgiu
ro014c	Alba Iulia
se001c	Stockholm
se002c	Göteborg
se003c	Malmö
se004c	Jönköping
se005c	Umeå
si001c	Ljubljana
si002c	Maribor
sk001c	Bratislava
sk002c	Kosice

Urban Audit code	City name
sk003c	Banska Bystrica
sk004c	Nitra
uk001c	London
uk002c	Birmingham
uk003c	Leeds
uk004c	Glasgow
uk005c	Bradford
uk006c	Liverpool
uk007c	Edinburgh
uk008c	Manchester
uk009c	Cardiff
uk010c	Sheffield
uk011c	Bristol
uk012c	Belfast
uk013c	Newcastle upon Tyne
uk014c	Leicester
uk015c	Derry
uk016c	Aberdeen
uk017c	Cambridge
uk018c	Exeter
uk019c	Lincoln
uk020c	Gravesham
uk021c	Stevenage
uk022c	Wrexham
uk023c	Portsmouth
uk024c	Worcester

Notes

1 This Google search was carried out on 8 April 2009.
2 PPP methods make it possible to better represent spatial disparities in the level of prices, and, consequently, more accurately gauge the real spending power of economic agents.
3 Del Bo and Florio (2012) offer a critical perspective on previous studies regarding the role of different forms of infrastructure in economic performance and provide empirical evidence on the contribution of single and aggregate measures of infrastructure on regional economic performance in the EU.
4 The use of italics in this list indicates a citation from Hollands (2008). On this first point, see also Komninos (2002).
5 Evidence of this last finding is available from the authors upon request.

References

Abreu, M., Grinevich, V., Kitson, M. and Savona, M. (2008) *Absorptive Capacity and Regional Patterns of Innovation*, Research Report DIUS RR-08–11, Cambridge, MA: MIT.

Arrow, K.J. (1962) The Economic Implications of Learning By Doing, *Review of Economic Studies*, 29: 155–177.

Berry, C.R. and Glaeser, E.L. (2005) The Divergence of Human Capital Levels Across Cities, *Papers in Regional Science*, 84 (3): 407–444.

Breschi, S. and Lissoni, F. (2001) Localised Knowledge Spillovers vs. Innovative Milieux: Knowledge 'Tacitness' Reconsidered, *Papers in Regional Science*, 80 (3): 255–273.

Bronstein, Z. (2009) Industry and Smart City, *Dissent*, 56 (3): 27–34.

Bush, V. (1945) *Science: The Endless Frontier*, Washington DC: United States Government Printing Office.

Capello, R. (2009) Spatial Spillovers and Regional Growth: A Cognitive Approach, *European Planning Studies*, 17 (5): 639–658.

Caragliu, A. and Nijkamp, P. (2012) The Impact of Regional Absorptive Capacity on Spatial Knowledge Spillovers, *Applied Economics*, 44 (11): 1363–1374.

Coe, A., Paquet, G. and Roy, J. (2001) E-governance and Smart Communities: A Social Learning Challenge, *Social Science Computer Review*, 19 (1): 80–93.

Cohen W. and Levinthal, D. (1990) Absorptive Capacity: A New Perspective on Learning and Innovation, *Administrative Science Quarterly*, 35 (1): 128–152.

Davis, D.R. and Weinstein, D.E. (2003) Market Access, Economic Geography and Comparative Advantage: An Empirical Test, *Journal of International Economics*, 59 (1): 1–23.

Del Bo, C. and Florio, M. (2012) Infrastructure and Growth in a Spatial Framework: Evidence from the EU Regions, *European Planning Studies*, 20 (8): 1393–1414.

Florida, R.L. (2002) *The Rise of the Creative Class: And How It's Transforming Work, Leisure, Community and Everyday Life*, New York: Basic Books.

Florida, R.L. (2009) Class and Well-being, www.creativeclass.com/creative_class/2009/03/17/class-and-well-being/, accessed 17 March 2009.

Fu, S. (2007) Smart Café Cities: Testing Human Capital Externalities in the Boston Metropolitan Area, *Journal of Urban Economics*, 61 (1): 86–111.

Giffinger, R., Fertner, C., Kramar, H., Kalasek, R.. Pichler-Milanović, N. and Meijers, E. (2007) Smart Cities: Ranking of European Medium-sized Cities, Final Report (October 2007), www.smart-cities.eu/download/smart_cities_final_report.pdf.

Glaeser, E.L. (2005) A Review of Richard Florida's 'The Rise of the Creative Class', *Regional Science and Urban Economics*, 35: 593–596.

Glaeser, E.L. and Berry, C.R. (2006) *Why Are Smart Places Getting Smarter? Taubman Cente Policy Brief 2006–2*, Cambridge, MA: Taubman Centre.

Graham, S. and Marvin, S. (1996) *Telecommunications and the City: Electronic Spaces, Urban Place*, London: Routledge.

Hanson, G.H. (2005) Market Potential, Increasing Returns and Geographic Concentration, *Journal of International Economics*, 67 (1): 1–24.

Hollands, R.G.(2008) Will the Real Smart City Please Stand Up?, *City*, 12 (3): 303–320.

Komninos, N. (2002) *Intelligent Cities: Innovation, Knowledge Systems and Digital Spaces*, London: Spon Press.

Lucas, R.E. (1988) On the Mechanics of Economic Development, *Journal of Monetary Economics*, 22: 3–42.

McAllister, L.M. Hall, H.M., Partridge, H.L. and Hallam, G.C. (2005) Effecting Social Change in the 'Smart City': The West End Connect Community Project, in C. Bailey and K. Barnett, eds, *Social Change in the 21st Century*, 28 October 2005, Brisbane, Australia.

Mahizhnan, A. (1999) Smart Cities: The Singapore Case, *Cities*, 16 (1): 13–18.

Mankiw, N.G., Romer, D. and Weil, D.N. (1992) A Contribution to the Empirics of Economic Growth, *The Quarterly Journal of Economics*, 107 (2): 407–437.

Mayer, H. (2007) What is the Role of the University in Creating a High-technology Region?, *Journal of Urban Technology*, 14 (3): 33–58.

Nijkamp, P. (2010) E pluribus unum, *Region Direct*, 2 (2): 56–65.

OECD – EUROSTAT (2005) *Oslo Manual*, Paris: Organization for Economic Cooperation and Development – Statistical Office of the European Communities.

Partridge, H.L. (2004) Developing a Human Perspective to the Digital Divide in the 'Smart City, *Australian Library and Information Association Biennial Conference*, 21–24 September 2004, Gold Coast, Queensland, Australia.

Poelhekke, S. (2006) Do Amenities and Diversity Encourage City Growth? A Link Through Skilled Labor, *Economics Working Papers ECO2006/10*, San Domenico di Fiesole, Italy: European University Institute.

Rauch, J.E. (1993) Productivity Gains from Geographic Concentration of Human Capital: Evidence from the Cities, *Journal of Urban Economics*, 34 (3): 380–400.

Redding, S.J. and Sturm, D.M. (2008) The Costs of Remoteness: Evidence from German Division and Reunification, *The American Economic Review*, 98 (5): 1766–1797.

Roller L-H. and Waverman, L. (2001) Telecommunication Infrastructure and Economic Development: A Simultaneous Approach, *American Economic Review*, 91 (4): 909–923.

Science Museum (2004) Urban Development, www.makingthemodernworld.org.uk/learning_modules/geography/04.TU.01/?section=2, accessed 3 April 2009.

Shapiro, J.M. (2008) Smart Cities: Quality of Life, Productivity, and the Growth Effects of Human Capital, *The Review of Economics and Statistics*, 88 (2): 324–335.

Southampton City Council (2006) Southampton On-line, www.southampton.gov.uk/thecouncil/thecouncil/you-and-council/smartcities/, accessed 13 March 2009.

United Nations (2009) World Urbanization Prospects: The 2009 Revision, http://esa.un.org/unpd/wup/index.htm, accessed 15 February 2011.

11 An advanced triple helix network framework for smart cities performance

Karima Kourtit, Mark Deakin, Andrea Caragliu, Chiara Del Bo, Peter Nijkamp, Patrizia Lombardi and Silvia Giordano

Introduction

Focusing on a subset of European cities belonging to the SmartCities (inter) Regional Academic Network (SCRAN), i.e. Bremerhaven, Edinburgh, Groningen, Karlstad, Kortijk, Kristiansand, Lillesand, Osterholz and Norfolk County, this chapter offers a decision network model built around an analytical hierachy able to verify whether the development of cities within the North Sea region is smart. It offers an in-depth analysis of the interrelations between the components of smart cities, including the human and social relations connecting the intellectual capital, wealth and governance of their regional development. This serves to demonstrate how the inclusion of the abovementioned relations in an analytical hierarchy framework makes it possible for the network model in question to capture the advanced triple helix of smart cities and verify whether the transformation this ushers in is based not only on the creation of wealth, but also on the governance of their underlying regional innovation systems.

The increasing dominance of cities

The past centuries have shown the increasing dominance of cities in the global landscape. It is not, however, the sheer population members that count, but the functional leadership of cities in a modern world. Cities are not just geographical settlements of people, they are also the 'home of man'. They reflect the varied history of mankind and are at the same time contemporaneous expressions of the diversity of human responses to future challenges. An example of the way urban architecture reflects and shapes the future can be found in Dubai, a city that has deliberately left behind its old history and decided to generate a spectacular new vision of itself as smart.

Dubai is not an exception, but may be seen as a trend setter. Over the past decade modern urban planning has promoted a multitude of initiatives centred around the creativity of urban development and on cultural artefacts as the cornerstones of such actions. Consequently, it has become fashionable to regard the types of creative expressions found in the cultural sector as the sign of things to come (see Florida, 2002). Under this process of urban development, 'old' cities such as London, Liverpool, Amsterdam, Berlin, Barcelona, New York, San Francisco, Sydney and Hong Kong are undergoing a profound transformation based on the rise of a creative sector. This new orientation does not only provide a new dynamism, it also has a symbolic value by showing the historical strength of cities as foundation stones for a new and open future.

Since Florida's (2002) ideas on the creative class, industry and city, a number of studies has been undertaken to examine the features and success conditions of creative environments (see, for example, Fusco Girard et al., 2009; Gabe, 2006; Heilbrun and Gray, 1993; Hesmondhalgh, 2002; Landry, 2001; Markusen, 2006; Power and Scott, 2004; Pratt, 1997; Vogel, 2001). Despite several empirical studies, however, an operational conceptualisation of creative infrastructures and the relation with the smart city concept has as yet not been developed and calls certainly for more focused and targeted research.

In responding to this call, the authors of this chapter adopt a recent conceptualisation and consider a city as smart: 'when investments in human and social capital and traditional (transport) and modern (ICT) communication infrastructure fuel sustainable economic growth and a high quality of life, with a wise management of natural resources, through participatory governance' (Caragliu et al., 2011: 70 and Chapter 12, this book). Furthermore, this chapter suggests that cities can become 'smart' when underlying investment by universities, industry and government supports the development of communication infrastructures and this in turn fuels sustainable economic growth and a high quality of life.

In moving towards such a representation of smart cities, this chapter will take on the notion of the triple helix as its starting point. As the main exponents of the triple helix, Etzkowitz and Leydesdorff (2000) offer a particularly insightful critique of so-called Mode 2 accounts of innovation, but limit their representation of the model to those institutional relations surrounding university, industry and government involvement in the knowledge economy of regional systems. In this literature, the focus is on the process of knowledge production, which is conceived as being mainly carried out in universities and firms, and whose officialisation (i.e. the process of patenting novel knowledge) is regulated according to standards laid down by governments.

While offering many critical insights into the political economy of the triple helix, it is noticeable that traditional triple helix studies reveal little about either the social basis of university, industry and government involvement, or the technical infrastructures of their regional innovation systems (Deakin, 2010: 4).This is because such studies of the triple helix fail to realise that universities, industry and governments only understand each other when the social and intellectual soil is fertile enough for knowledge to flow between them. The blind spot in such representations of the triple helix can no doubt be explained by their tendency to study the interrelations of university and industry both apart from, in isolation to and, it might be added, at the expense of, their governance by institutions within the public realm.

The authors of this chapter should like to suggest that this blind spot in the traditional representations of the triple helix is significant because the oversight reveals that the current model's concern lies with university and industry relations and not the governance they are subjected to as part of the laying down of standards put in place by the government axis to regulate their development. The problem with this position is that in western democratic societies it is not just universities or industry who invest in technical infrastructures, but government which also gets involved in bank rolling such ventures. Governments act not only as investors but also as those institutions responsible for setting the conditions that influence the final return that such ventures may generate.

As shall be seen, this chapter's decision to include contour conditions, and especially relations between the three traditional helices, within the original triple helix, offers a series of critical insights into the transformation that cities are currently experiencing as part of their drive to become smart, and offers fertile ground for those in the university and industrial sectors seeking to magnify returns derived from the types of investments underpinning the technical infrastructures of their regional innovation systems. In mapping out the contour of this landscape, the advanced triple helix offers a more detailed analysis of the interrelations between smart city components. In order to achieve such a goal, this chapter provides an extension of the traditional triple helix model by focusing both on the hard (physical, infrastructure-related) components of the triple helix and the soft (non-material, intangible) growth-enhancing characteristics of the analysed cities. In particular, our analysis is extended to the wealth of human, social, environmental and economic relations of the intellectual capital that forms the fertile soil for a smart city. This analysis of the triple helix shall then be augmented by using the Analytic Network Process (ANP) provided by Saaty (2005) as a basis to model, cluster and begin measuring their performance.

Towards smart cities

Recent debates on smart cities have tended to shift away from the high-tech architecture that has come to symbolise the subject and towards the economic development, environment, human and social capital, culture, leisure and governance by which it is increasingly becoming known. This is captured by the recent introduction of soft factors, such as 'smart economy', 'smart mobility', 'smart environment', 'smart people', 'smart living' and 'smart governance', into the syntax of smart cities (www.smart-cities.eu).

This chapter specifically focuses on the role of digital infrastructures and information and communication technologies (ICTs) in supporting the governance of smart cities and the measures some cities are taking to be smart in meeting the regional innovation challenge this poses. The analysis is based on an extensive study of the response made to this challenge by European countries and regions bordering the North Sea and as evidenced by the digital technologies they have assembled to support their governments' development of electronically enhanced services. The European cities involved in this exercise are those belonging to the SmartCities (inter) Regional Academic Network (SCRAN), i.e. Bremerhaven, Edinburgh, Groningen, Karlstad, Kortijk, Kristiansand, Lillesand, Osterholz and Norfolk (Deakin, 2010, 2011). Since the development of these digital infrastructures draws extensively upon the capacities of universities and industry, as much as the regulative powers of governments, the case study also offers the opportunity to test whether the limitations of traditional triple helix accounts can be overcome by the present analysis run by SCRAN and extended to cover governance issues as well, in particular on themes such as the regulation of the provision of electronically enhanced services (eGov).

Digital infrastructure, ICTs and the development of eGov services

It is often claimed that some cities in Europe are smart in using the digital infrastructures of ICTs to develop eGov services. Claims made about their use of ICTs to innovate and develop eGov services as online services testify to this. Recent surveys of these developments, however, also serve to raise a number of questions about whether such ICT-driven innovations are smart and if cities should be creating opportunities for online services offering 24/7 access.

As Torres et al. (2005) have noted, the absence of any commonly agreed terminology to describe the technical infrastructures of such developments has left policy makers without the means to discuss such matters and to agree on what they represent. In an attempt to overcome this and begin to understand what such developments mean, Torres et al. (2005) and others (for example, Lee et al., 2006; Lombardi and Cooper, 2009; Deakin, 2009) have sought to develop a user-centric and customer-focused terminology capable of supporting a standard classification of the eGov service developments in question (Deakin, 2010). Having made headway with this standardisation of eGov service developments, attention has now begun to turn towards the question of what it means for cities to be smart. Others have taken a more 'measured' approach and sought to use the indicator sets currently available as the statistical base for analysing what smart city developments of this kind contribute to sustainable urban development (Torres et al., 2005; Caragliu et al., 2011; and Chapter 10, this book). However, answering the question of what it means for cities to be smart has also been hampered, not so much by the need to agree on a standard representation of eGov service developments, but by the lack of a robust statistical base to measure them by. For some this has resulted in a desperate cry for the 'real smart city to stand up', to explain what it means for a city to be smart and how the eGov service developments embarked upon to gain such a standing can be measured (Hollands, 2008; Deakin, 2010; Caragliu et al., 2011; and Chapter 10 of this book).

In responding to this cry for the 'real smart city to stand up' and be analysed, the authors of this chapter advance an extended triple helix to baseline the development of smart cities in terms of their traditional and contemporary roles: first as generators of intellectual capital, creators of wealth and regulators of standards (university, industry and government, respectively), then as cities that use such attributes to be smart in supporting the social learning, market-based entrepreneurial capacities and knowledge-transfer abilities that are needed to meet the requirements of their regional innovation systems. Reporting on the outcomes of this evaluation, the chapter argues that such assessments are particularly important because the work published by Torres et al. (2005) has defined those authorities responsible for promoting eGov service developments as 'steady achievers', not particularly innovative or creative. This in turn tends to suggest that any evaluation of the developments under way within the North Sea Region needs first of all to discover what it is that makes them steady achievers when compared with the more innovative, creative, namely, smarter counterparts in mainland and southern Europe. Then to draw upon the outcome of this transnational assessment to uncover what innovation is required in order for smart cities to develop across the North Sea Region.

Smart city metrics

The triple helix model has emerged as a reference framework for the analysis of knowledge-based innovation systems, and charts the multiple and reciprocal relationships between the three main agencies in the process of knowledge creation and capitalisation: university, industry and government (see for a recent overview Etzkowitz, 2008). In the context of the present analysis, we shall focus on this model as a starting point for assessing the performance of smart cities (Leydesdorff and Deakin, 2011; Deakin, 2011; 2012; and Chapter 10 of this book). In order to link the evaluation of smart city components and the three main helices of the model, we propose a modified triple helix framework by adding another unifying factor to the analysis, namely urban environments and their contour conditions.

While it is accepted by the authors of this chapter that knowledge is created by the interplay of the relations captured in the original triple helix, with the advanced triple helix model it is proposed that the accumulation of capital within regional innovation systems is enhanced by way of interaction with urban environments and through their contour conditions. Contour conditions that not only contribute to the generation of intellectual capital, or to the creation of wealth within smart cities, but also to the standards that government draws upon to regulate the accumulation of capital as wealth within regional innovation systems.[1] In this way, the advanced triple helix model provides the hitherto missing link between urban innovation systems and smart

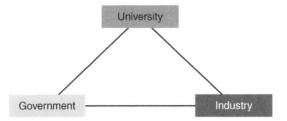

Figure 11.1 The original triple helix

cities. This is because it presupposes that the three helices operate within the landscape of a complex urban environment, where market demand, governance, civic involvement and citizenship, along with their cultural and social endowments, not only shape the relationships between the traditional helices of university, industry and government, but of the regional innovation system that they advance.

Our framework can be exemplified by the following figure (Figure 11.1) and can be operationalised by focusing on the measurement of the three main helices and the contour conditions and linking these to a smart city indicator.

The original triple helix analyses the crucial role of the interplay between the three main helices of the innovation system, namely university, industry and government (see Figure 11.1). However, in this model scarce attention is paid to the output generated by, and the filters intervening in, the relations between each of the traditional triple helix axes. In this chapter we suggest a modified version of the triple helix (Figure 11.2), where such filters are evidenced and seen as contributing to our understanding of the urban and regional innovation systems in question. In fact, we believe that the efficiency of the knowledge exchange among the actors of the triple helix is enhanced both by way of and through the filters suggested in our model. In particular:

- the knowledge stock generated by the interplay between universities and industry contributes to the generation of trustworthy relations between the two, and represents an asset for future learning performance (Cohen and Levinthal, 1990);
- collective learning mechanisms take place, as universities and government bodies act together in searching for efficient public management solutions, causing the creative resonance mechanism at the basis of innovation processes;
- finally, the 'thickness' and efficiency of market institutions and actors are strongly related to the efficiency with which industry and government exchange information and generate innovative products and processes. This is probably the most strikingly absent element within the original triple helix approach, whereas the absence of market institutions

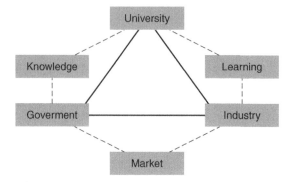

Figure 11.2 The advanced triple helix

and relations represents a crucial missing link in explaining innovation processes.

According to this scheme:

- Knowledge is the result of the interaction between university and industry. In fact, as pointed out in Etkowitz and Leydesdorff (2002), the European Innovation System is relatively lagging in this respect, while laws fostering such interactions, as the Bayh-Dole Act in the US, may provide positive incentives for establishing successful research activities.
- The interplay between university and government produces what is here labeled as 'learning'. Here public institutions learn from educational bodies ways to improve their performance and take advantage of a better educated workforce. At the same time, the university system benefits from an efficient management of public goods. The mutual reinforcement of this mechanism generates society-level learning.
- An efficient market, based on well-defined rules and functioning institutions, not only guarantees cooperation between the private and government sectors, but enhances the interrelations between universities, industry and government at the local level where knowledge is produced.

It is these three elements, namely knowledge, learning and their institutionalisation within the social contours of the market, that represent the innovative component of our approach and form the means by which to magnify the returns of the original triple helix.

We adopt this original framework in order to analyse cities belonging to the SCRAN network, and assess the connections between smart city development and this institutionalisation of the advanced triple helix. As pointed out above, traditional smart city definitions usually focus on digital services. While the original triple helix already suggests the need for a broader perspective, here

Table 11.1 Advanced triple helix data

Context	Element	Measure
Original Triple Helix	University	University (% people aged 20–24 enrolled in tertiary education)
Original Triple Helix	Industry	Industry (Number of companies per 1,000 pop.)
Original Triple Helix	Government	Government (% labour force in government sector L to Q: public administration and community services; activities of households; extra-territorial organisations)
Advanced Triple Helix	Learning	Learning (labour force with ISCED 5 and 6 education)
Advanced Triple Helix	Market	Market (Per capita GDP)
Advanced Triple Helix	Knowledge	Knowledge (Patent applications to the EPO per 1,000 inh.)

we claim that this can be further improved by considering the contour conditions in each city. It is for this reason that we analyse smart cities through such an institutionally grounded representation of the triple helix.

In undertaking this analysis, the research team has assembled a new data set, collecting information from the SCRAN cities. Among the data obtained, we identified one indicator suitable for each of the elements of the advanced triple helix. The data is described in Table 11.1.

This set of data is used to graphically represent the overall dimension of the advanced triple helix for the smart cities, in comparison with the average EU27 situation. This is illustrated in Figure 11.3.

With respect to the traditional three helices, smart cities are above the EU average. In fact, there is a higher percentage of young adults engaged in higher education, a higher share of labour force in the government sector and a higher intensity of firms per population. However, while smart cities match the average EU performance in terms of the market, they slightly under-perform in the knowledge and learning domain when compared with the EU average. This allows us to conclude that, while the cities in our sample are indeed moving in the right direction, there is still room for improvement. The lag in terms of contour conditions may hamper the positive and rich endowment with traditional triple helix elements.

Towards a more sound measure of 'smartness'

Previously, Caragliu et al. (2011 and Chapter 10 of this book) have stated that traditional definitions of smart cities fall short of their ambition because they tend to ignore contour conditions around urban digital characteristics. What follows suggests these contours can be mapped out by assembling indicators

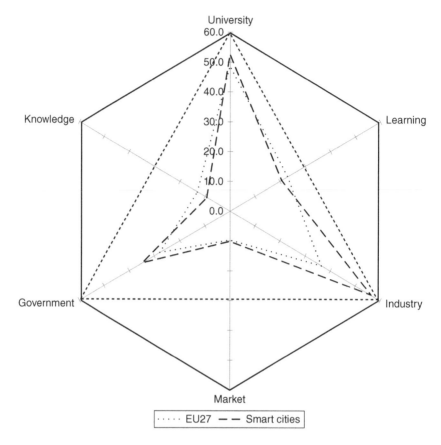

Figure 11.3 Advanced triple helix for the smart cities

on each of the six axes identified as critical in defining urban smartness. These indicators have been collected for each member of the SCRAN network and are the following: Percentage of households with internet access at home; Proportion of households with broad band access; Length of public transport network; Proportion of population aged 15–64 with some college education living in Urban Audit cities; Green space (in m²) to which the public has access, per capita; and Annual expenditure of the municipal authority per resident. Spatial variance in terms of these indicators is then summarised via a principal component analysis (PCA), whose first component (explaining 40 per cent of total variance) is labelled smartness.

PCA is a multivariate statistical technique aimed at identifying patterns in data and eventually compressing them by reducing the number of dimensions, each of which, orthogonal to the previous component, is the single subcomponent maximising the original variance in the data. This process has

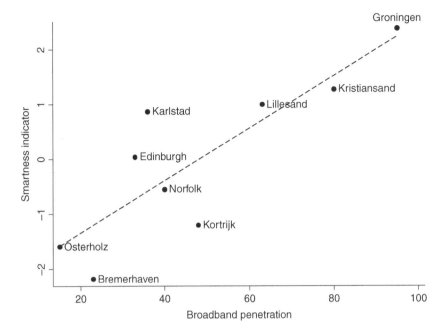

Figure 11.4 ICT penetration and smartness

the advantage of reporting the amount of variance in the data explained by each aggregate index. In practice, the original data is standardised, the covariance matrix is calculated and eigenvectors and eigenvalues are computed. Eigenvectors are then ordered with respect to associated eigenvalues, from highest to lowest. The principal components, or PC (i.e. eigenvectors with the highest eigenvalues, which are linear combinations of the original variables) are then selected according to the Jollife-amended Kaiser eigenvalue criterion and examination of the proportion of variance accounted by the principal components. Besides, all components are built in order to be orthogonal to each other.

Figure 11.4 relates the resulting aggregate indicator of smart city characteristics with ICT penetration and suggests that indeed smartness is strongly related to urban digital infrastructures. In Figure 11.4, the x-axis shows the rate of broadband penetration in smart cities, and the y-axis the smartness indicator already described. The two indicators have a Pearson's correlation index of 0.84, significant at all conventional levels. This result suggests that a broader definition of smart city, while on the one hand extending traditional narrow definitions of urban smartness from digital technologies to a more comprehensive view of such services, is still significantly linked to ICTs.

As noted earlier, the success of urban and regional innovation systems, represented in the advanced triple helix model, does not solely depend on the

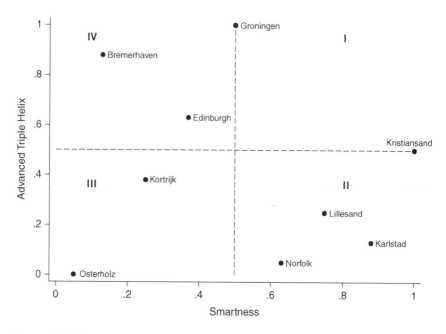

Figure 11.5 Smart cities and the advanced triple helix

technical infrastructures of ICTs, but on the broadband of both the hard and soft characteristics of their environment. Within the advanced triple helix, smartness and good positioning in terms of the revised triple helix do not necessarily coincide. In order to make this statement evident we have built an indicator of performance within the advanced triple helix based on the PCA of the six indicators built to assess the smartness of cities. The first component explains around 40 per cent of total variance in the data.

In Figure 11.5 we plot the smartness indicator (described above) on the x-axis and the revised triple helix index on the y-axis for nine SCRAN cities. The graph shows a vertical and a horizontal dashed line, corresponding to half distribution in terms of both indicators. In this way we identify four quadrants (first quadrant on the top right of the graph, to be read clockwise).

The first striking result is that no city scores high with respect to both indicators, highlighting a potential direction for future improvement. In quadrant II we observe cities scoring high in terms of ICT endowment, but relatively worse in terms of structural innovation-oriented characteristics. In quadrant IV the opposite happens, with cities showing a good performance of traditional triple helix elements, but less rich in terms of ICTs. Quadrant III, finally, shows two cities with potential for improvement along both dimensions. Notice that the graph is built solely on SCRAN data; as such, it tells us nothing about their relative positioning with respect to potential competitors.

In fact, as Table 11.3 clearly shows, SCRAN cities score relatively high with respect to the EU average in most triple helix indicators.

The analysis presented in this section shows some potential pitfalls of identifying smartness solely with the ICT dimensions, and links the smart city framework with the advanced triple helix, showing the potential for cross-fertilisation of the two areas of research. Empirical results also highlight the need and stress the importance of analysing several dimensions of the urban environment in order to assess a city performance.

Additional research directions

The results of the study illustrated in the previous sections suggest how to baseline the development of smart cities in terms of their traditional and contemporary roles: first as 'urban innovators' in the generation of intellectual capital, creators of wealth and regulators of standards (university, industry and government, respectively), then as cities that need such attributes to be smart in order to support the social learning, market-based entrepreneurial capacities and knowledge-transfer abilities that meet the requirements of their regional innovation systems.

Further research currently being undertaken aims to establish the steps that cities are taking to deploy the ICTs of e-government as a means to get beyond the baseline. In particular we focus on the steps they are taking to adopt ICTs as a means to develop their public services in line with the needs of the information society and the requirements this in turn places on the knowledge economy of smart cities. This extended analysis will also serve to highlight some of the 'rich ecologies' that underpin the environmental sustainability of these service developments and upon which the smartness of their information society and knowledge economy in turn rest. Here the amount of green space, that is the space dedicated to leisure and recreational use and the length of the public transportation indicators, has been selected and used as a means to indicate how biodiversity and carbon reduction measures may not only be linked to social mobility, but also connected to an overall concept of smartness.

Good performance along these dimensions indicates that city planners have taken steps towards greater sustainability and value not only technological and knowledge-related aspects of smart urban evolution, but are concerned with quality of life and aim for cities to become green. Figure 11.6 shows the performance of smart cities in terms of eco-sustainability with respect to the urban European average. The authors of this chapter have considered the square meters of green area each inhabitant has access to, the length in kilometres of public transportation per square kilometre and the percentage area used for recreational purposes. As can be inferred from a visual inspection of the graph, smart cities can also be labelled as green cities, since areas devoted to recreation and green are above the European average, and the public transportation network's length is comparable to the other EU cities.

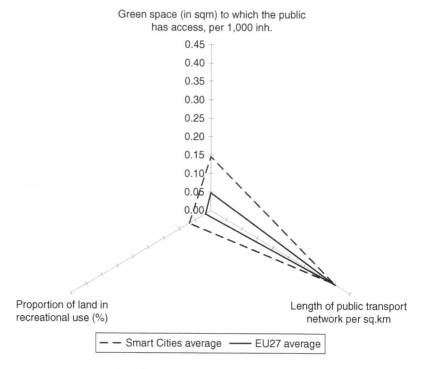

Figure 11.6 An evaluation of smart cities' performance in terms of eco-sustainability with respect to the urban European average

This study is ongoing and currently authors are exploring the possibility of identifying and measuring the relations between the smart cities components, including smart governance, smart human capital, smart environment, smart living and economy, by way of and through an ANP model. The ANP is the first mathematical theory that makes it possible to systematically deal with all kinds of dependencies (Saaty, 2005). This is helpful for highlighting the relationships between triple helix indicators and smart cities components, as highlighted in Table 11.2. As one can notice, in this application, the triple helix includes also the civil society as a key player in the sustainable urban development agenda (Etzkowitz and Zhou, 2006; Lombardi and Cooper, 2009; Brandon and Lombardi, 2011; Lombardi et al., 2012).

The source of the data for this table are both a literature review, including EU projects' reports and Urban Audit dataset, and indicators selected from statistics of the European Commission, European green city index, TISSUE, Trends and Indicators for Monitoring the EU Thematic Strategy on Sustainable Development of Urban Environment and smart cities' ranking of European medium-sized cities.

Table 11.2 Smart cities' components and performance indicators

	SMART Governance	SMART Economy	SMART Human	SMART Living	SMART Environment
UNIVERSITY	Number of universities, research centres in the city	Public expenditure on R&D - % of GDP per head of city	% of population aged 15-64 with secondary level education (Urban Audit)	% of professors & researchers involved in international projects and exchange	An assessment of the ambitiousness of CO_2 emissions reduction strategy
	Number of courses entirely downloadable from the internet / Total number of courses	Public expenditure on education - % of GDP per head of city	% of population aged 15-64 with high education (Urban Audit)	Number of grants for international mobility per year	An assessment of the extensiveness of city energy efficiency standards for buildings
		Number of research grants funded by international projects	% of inhabitants working in education and in research & development sector	% of accessible courses	
GOVERNMENT	E-government on-line availability (% of the 20 basic services which are fully available online)	Gross Domestic Product per head	Voter turnout in national and EU parliamentary elections	Proportion of the area in recreational sports and leisure use	Total annual energy consumption in gigajoules per head
		Debt of municipal authority per inhabitant	Share of female city representatives	Green space (m²) to which the public has access, per capita	Efficient use of electricity (use per GDP)
	Percentage of households with computers	Median or average disposable annual household income	City representatives per resident	Number of public libraries	Total annual water consumption, in cubic metres per head
		Unemployment rate		Number of theatres & cinemas	Efficient use of water (use per GDP)
	Percentage of households with Internet access at home	Energy intensity of the economy - Gross inland consumption of energy divided by GDP		Health care expenditure - % of GDP per head of city	Area in green space (m²)
					Greenhouse gas emissions and intensity of energy consumption
				Tourist overnight stays in registered accommodation per year per resident	An assessment of the comprehensiveness of policies to contain the urban sprawl and to improve and monitor environmental performance

Table 11.2 (cont.)

					Urban population exposure to air pollution by particulate matter - micrograms per cubic metre
CIVIL SOCIETY	E-government usage by individuals (% individuals aged 16 to 74 who have used the Internet, in the last 3 months, for interaction with public authorities)	% of projects funded by civil society	Foreign language skills	Total book loans and other media per resident	The total percentage of the working population travelling to work on public transport, by bicycle and by foot
			Participation in life-long learning (%)	Museum visits per inhabitant	An assessment of the extent to which citizens may participate in environmental decision-making
			Individuals' level of computer skills	Theatre & cinema attendance per inhabitant	An assessment of the extensiveness of efforts to increase the use of cleaner transport
			Individuals' level of internet skills		% of citizens engaged in environmental and sustainability oriented activity
INDUSTRY	Number of research grants funded by companies, foundations, institutes / No annual scholarships	Employment rate in: - High-tech & creative industries - Renewable energy & energy efficiency systems - Financial intermediation and business activities - Culture & entertainment industry - Commercial services - Transport and communication - Hotels and restaurants	Patent applications per inhabitant	Number of enterprises adopting ISO 14000 standards	The percentage of total energy derived from renewable sources, as a share of the city's total energy consumption
					Combined heat and power generation - % of gross electricity generation
		All companies (total number)	Employment rate in knowledge-intensive sectors	Rate of people undertaking industry based training	Rate of recycled waste per total kg of waste produced
		Number of local units manufacturing high-tech & ICT products			

		Total CO^2 emissions, in tonnes per capita (2)
Companies with HQ in the city quoted on national stock market		
Components of domestic material consumption		% of new buildings and renovations with sustainability certification

The assessment model requires a number of steps, starting from an identification of the relationships between the elements constituting the decision problem. There are two kinds of interdependences: one between elements (indicators) related to different clusters (external connection) and one within the same cluster (internal relation). The latter one is identified as a loop. Among the external connections, there are either mono-directional relationships, when one indicator is depending on another, or bidirectional relationships, when the dependency between indicators is reciprocal.

The subsequent step is the development of a pair-wise comparison of both elements and clusters to establish relations within the structure. In this step, a series of pair-wise comparisons are made by participants in the decision-making process (usually experts, managers and citizen representatives) to establish the relative importance of decision elements with respect to each component of the network. In pair-wise comparisons, a ratio scale of 1–9 numbers is used (called the fundamental scale, or Saaty scale). The numerical judgements established at each level of the network form pair matrixes which are used to derive weighted priority vectors of elements (Saaty, 2001).

Finally, to obtain the global priority vector of the elements, including the alternatives, the mathematical approach encompasses the use of super-matrices. The super-matrix which contains the global priority vector, i.e. a long-term stable set of weights, is obtained by rising to limiting power the weighted super-matrix.

An exercise of this kind has been recently developed by Lombardi et al. (2012) and Lombardi and Giordano (2012) with the aim of evaluating four different policy visions of the city of the future as derived from the 'Urban Europe' Joint Programme Initiatives (Nijkamp and Kourtik, 2011):

The Connected City (smart logistic and sustainable mobility). The image of a connected city refers to the fact that in an interlinked (from local to global) world, cities can no longer be economic islands in themselves ('no fortresses'), but have to seek their development opportunities in the development of advanced transportation infrastructures, smart logistic systems and accessible communication systems through which cities become nodes or hubs in polycentric networks (including knowledge and innovation networks).

The Entrepreneurial City (economic vitality). This vision assumes that in the current and future global and local competition, Europe can only survive, if it is able to maximise its innovative and creative potential in order to gain access to emerging markets outside Europe; cities are then spearheads of Europe's globalisation policy.

The Pioneer City (social participation and social capital). This vision refers to the innovative 'melting pot' character of urban areas in the future, which will show an unprecedented cultural diversity and fragmentation of lifestyles in European cities; this will prompt not only big challenges, but also great opportunities for smart and creative initiatives in future cities, through which Europe can become a global pioneer.

The Liveable City (ecological sustainability). The final vision addresses the view that cites are not only energy consumers (and hence environmental polluters), but may – through smart environmental and energy initiatives (e.g. recycling, waste recuperation) – act as engines for ecologically benign strategies, so that cities may become climate-neutral agents in a future space-economy; cities in Europe are then attractive places to live and work.

The results show that the *Entrepreneurial City* is the policy vision with higher priorities in all the sectors considered in the model, i.e. universities, government, civil society and industry. This means that a high degree of entrepreneurial activities and a constant flow of new firm creation is a prerequisite for finding a new role within the new global economic landscape. Innovation and creativeness are thus the necessary ingredients for entrepreneurial cities in Europe.

Although the proposed evaluation model and pilot exercise still requires testing and further application with the participation of real city stakeholders, it offers a reflexive learning opportunity for the cities to measure what options exist to improve their performances. Some relevant urban planning and policy implications of this vision are provided as follows.

- A high degree of entrepreneurial activities and a constant flow of new firm creation is a prerequisite for finding a role within the new global economic landscape. Innovation and creativity are thus the necessary ingredients for entrepreneurial cities in Europe.
- Special emphasis has to be given to new architectures, building technologies, intra-urban mobility solutions, public space management, e.g. for lighting or citizen information management, integrated urban energy planning and management and ICT-based solutions which offer various opportunities for new urban design and management.
- New requirements for efficient, effective and reliable infrastructures (such as energy, ICT, water, waste treatment and management) may occur. Since an appropriate infrastructure is essential for a city's attractiveness to companies and people and therefore to their economic development,

emphasis has to be given to the determination of these requirements within the scope of cities as complex systems.

Conclusions

As briefly illustrated in the introduction to this chapter, smart city concepts have recently become rather fashionable. A main challenge of the modern creative fashion in smart cities is to translate creative and cultural assets and expressions into commercial values (value added, employment, visitors, etc.), which means that private-sector initiatives are a sine qua non for effective and successful urban innovation strategies. Consequently, an orientation towards local identity and local roots ('the sense of place'), a prominent commitment of economic stakeholders (in particular, the private sector) and the creation of a balanced and appealing portfolio of mutually complementary urban activities are critical success conditions for a flourishing urban innovation strategy.

Smart cities offer, through their agglomeration advantages, a broad array of business opportunities for creative cultures, in which in particular self-employment opportunities and small- and medium-sized enterprises (SMEs) may play a central role in creating new urban vitality. Clearly, flanking and supporting urban conditions, such as local identity, an open and attractive urban 'milieu' or atmosphere, usage of tacit knowledge, presence of urban embeddedness of new business initiatives and access to social capital and networks provide additional opportunities for a booming urban creativeness culture and an innovative, vital and open urban social ecology. Urban creativeness presupposes an open and multi-faceted culture and policy supported by smart people.

This chapter has tried to answer the following question: if smart cities build the capacity for information society and support civic engagement by way of eGovernment services and through their development, what role do the institutions underlying all of this play in the process? In answering this question, the analysis has revised the triple helix concept, considering that universities, industry and government only understand each other when the social and intellectual ground linking them together is fertile enough to connect one with the other. These linkages and connections have been developed under the advanced triple helix model that this chapter has sought to outline. For in addition to university, industry and government relations set out in traditional representations of the triple helix, this model includes three other elements, namely knowledge, learning and their institutionalisation within the urban environment. The inclusion of contour and market conditions may in fact magnify returns derived from investments in the technical infrastructures of the regional innovation systems.

In order to support this advanced triple helix, a quantitative analysis has been developed, focusing on a subset of European cities belonging to the SCRAN network in comparison with the EU average. The results obtained stress the importance of analysing several dimensions of the urban environment in order to assess smart city performance and link it to the elements of the advanced triple helix. This is the reason why we have developed a further analytical step, considering four clusters of indicators, linked together and connected to one another in a network model. Here an analytical network framework has to be adopted, as it helps capture the triple helix of a smart urban development and verify whether the implied transformation of cities is based not merely on an index of intellectual capital, but on a measure of wealth creation whose standards of governance are set in the public realm.

Notes

1 In fact, this point refers to the notion of synergies among institutional actors in the triple helix and lies at the very heart of this literature.

References

Brandon, P.S. and Lombardi, P. (2011) *Evaluating Sustainable Development in the Built Environment*, 2nd edition, London, Wiley-Blackwell.
Caragliu, A., Del Bo, C. and Nijkamp, P. (2011) Smart Cities in Europe, *Journal of Urban Technology*, 18 (2): 65–82.
Cohen, W. and Levinthal, D. (1990) Absorptive Capacity: A New Perspective on Learning and Innovation, *Administrative Science Quarterly*, 35 (1): 128–152.
Deakin, M. (2009) A Community-based Approach to Sustainable Urban Regeneration, *Journal of Urban Technology*, 16 (1): 91–110.
Deakin, M. (2009) The IntelCities Community of Practice: The eGov Services Model for Socially-inclusive and Participatory Urban Regeneration Programmes, in Reddick, C., ed., *Research Strategies for eGovernment Service Adoption*, Hershey, Idea Group Publishing.
Deakin, M. (2010) SCRAN's Development of a Trans-national Comparator for the Standardization of e-Government Services, in Reddick, C., ed., *Comparative e-Government: An Examination of e-Government Across Countries*, Berlin, Springer Integrated Series in Information Systems.
Deakin, M. (2011) SCRAN: Assembling a Community of Practice for Standardizing the Transformation of eGovernment Services, *Networked Industries Quarterly*, 13 (3): 18–21.
Deakin, M. (2012) SCRAN: Assembling a Community of Practice for Standardizing the Transformation of eGovernment Services, in Aikins, S., ed., *Managing E-Government Projects: Concepts, Issues and Best Practices*, Hershey, ICI Publisher.
Etzkowitz, H. (2008) *The Triple Helix: University-Industry-Government Innovation in Action*, London, Routledge.
Etzkowitz, H. and Leydesdorff, L. (2000) The Dynamics of Innovation: From National Systems and 'Mode 2' to a Triple Helix of University–industry–government Relations, *Research Policy*, 29 (2): 109–123.
Etzkowitz, H. and Leydesdorff, L. (2002) *Universities and the Global Knowledge Economy NIP: A Triple Helix of University-industry-government Relations*, Continuum International Publishing Group Ltd.

Etzkowitz, H. and Zhou, C. (2006) Triple Helix Twins: Innovation and Sustainability, *Science and Public Policy*, 33 (1): 77–83.

Florida, R. (2002) *The Rise of the Creative Class ... and How It's Transforming Work, Leisure, Community and Everyday Life*, New York City, NY, Basic Books.

Fusco Girard, L., Lombardi, P. and Nijkamp, P. (2009) Creative Urban Design and Development (special issue), *International Journal of Services Technology and Management*, 13 (2/3): 111–115.

Gabe, T.M. (2006) Growth of Creative Occupations in U.S. Metropolitan Areas, *Growth and Change*, 7 (3): 396–415.

Giffinger, R., Fertner, C., Kramar, H., Meijers, E. and Pichler-Milanović, N. (2007) Ranking of European Medium-sized Cities, Final Report, Vienna.

Hesmondhalgh, D. (2002) *The Cultural Industries*, London, Sage.

Heilbrun, J. and Gray, C.M. (1993) *The Economics of Art and Culture*, New York, NY, Cambridge University Press.

Hollands, R. (2008) Will the Real Smart City Please Stand Up?, *City*, 3: 303–320.

Landry, C. (2001) *The Creative City*, London, Earthscan.

Lee, L., Tan, X. and Trimmi, S. (2006) Current Practices of Leading e-Government Countries, *Communications of the ACM*, 48 (10): 100–104.

Leydesdorff, L. and Deakin, M. (2011) The Triple Helix of Smart Cities: A Neo-evolutionist Perspective, *Journal of Urban Technology*, 18 (2): 53–63.

Lombardi, P. and Cooper, I. (2009) The Challenge of the eAgora Metrics, *International Journal of Services Technology and Management*, 13 (2/3): 210–222.

Lombardi, P. and Giordano, S. (2012) Evaluating the European Smart Cities Visions of the Future, *International Journal of the Analytic Hierarchy Process*, 4 (1), ISSN 1936–6744.

Lombardi P., Cooper, I., Paskaleva, K. and Deakin, M. (2009) The Challenge of Designing User-centric e-Services: European Dimensions, in C. Reddick, ed., *Strategies for Local E-Government Adoption and Implementation: Comparative Studies*, Hershey, Idea Group Publishing.

Lombardi, P., Giordano, S., Farouh, H. and Yousef, W. (2012) Modelling the Smart Cities Performances, *Innovation: The European Journal of Social Science Research*, 25 (2): 133–145.

Markusen, A. (2006) Urban Development and the Politics of a Creative Class, *Environment & Planning, A*, 38 (10): 1921–1940.

Nijkamp, P. and Cohen, G. (2010) Opportunities and Pitfalls of Local e-Democracy, in U. Allegretti, ed., *Democrazia Partecipativa*, Frienze, Firenze University Press: 200–215.

Nijkamp, P. and Kourtik, K. (2011) Joint Programming Initiative (JPI) on Urban Europe: Global Challenges and Local Responses in the Urban Century. *A Scoping Document*, 27, VU University, Amsterdam,.

Paskaleva-Shapira, K. (2008) Assessing Local e-Governance in Europe, *International Journal of Electronic Governance Research*, 4 (4): 17–36.

Power, D. and Scott, A. (2004) *Cultural Industries and the Production of Culture*, London, Routledge.

Pratt, A. (1997) The Cultural Industries Production System, *Environment & Planning A*, 29: 1953–1974.

Saaty, T. (2001) *The Analytic Network Process: Decision Making with Dependence and Feedback*, Pittsburgh, RWS Publications.

Saaty, T.L. (2005) *Theory and Applications of the Analytic Network Process*, Pittsburgh, RWS Publications.

Torres, L., Vicente, P. and Basilio, A. (2005) E-government Developments on Delivering Public Services Among EU Cities, *Government Information Quarterly*, 22: 217–238.

Van Soom, E. (2009) Measuring Levels of Supply and Demand for e-Services and e-Government: A Toolkit for Cities, *Smart Cities Research Brief*, N. 3, www.smartcities.info/research-briefs, accessed 25 February 2009.

Vogel, H. (2001) *Entertainment Industry Economics*, New York, NY, Cambridge University Press.

Websites

Smart Cities is a North Sea Intereg 4B project (2007–2013), www.northsearegion.eu/ivb/projects/details/&tid=84

Smart Cities – Ranking of European Medium-sized Cities, Centre of Regional Science, Vienna UT, October 2007, www.smart-cities.eu, accessed 25 November 2010.

Quality of Life in Twelve of New Zealand's Cities, Report 2007, www.qualityofifeproject.govt.nz, accessed 25 November 2010.

12 Conclusions (on the state of the transition)

Mark Deakin

Introduction

Taking Hollands' (2008) statement about the unspoken assumption as the object of enquiry, this book has reflected upon the anxieties currently surrounding the transition from intelligent to smart cities. In particular, his suggestion that such developments have more to do with cities meeting the corporate needs of marketing campaigns than the social intelligence required for them to be smart.

Working on the assumption that any attempt to overcome such anxieties means cities shifting attention away from the needs of the market and towards the intelligence required for them to be smart, the chapters in this book have begun to set out a less presumptious and more critically aware understanding of the transition. In particular, an understanding based on the legacy of research carried out on the informational basis of the communications that are embedded in such intelligence and which in turn leads society away from the purely technical issues underlying the business logic of those eGov service developments surfacing from the transition.

That is, away from the technical issues that surround the business logic of such developments and towards a more critically insightful appreciation of the information-rich and highly communicative qualities of the technologies supporting them. Away, in that sense, from the technical aspects of such developments and towards an examination of the social capital which is not only critical in underpinning their informational and communicative qualities, but also insightful in revealing the wider environmental and cultural role intelligence plays in supporting the transition to smart cities.

Armed with this critical insight, the book has sought to capture the highly communicative qualities of such service developments, the particular methodological issues that smart communities pose cities and the critically insightful role that the networks of innovation and creative partnerships set up to embed such intelligence play in the learning, knowledge transfer and capacity-building exercises servicing this community-led transition to smart cities.

This, it has been argued, is what Hollands' (2008) account of smart cities currently misses, and goes some way to explain why he asks 'the real smart city to stand up'. For in cutting across the social capital and environmental capacities of the transition from intelligent to smart cities, the representation that emerges is insufficiently grounded in the information systems and communications of the embedded intelligence which not only underlie such developments, but also surface as the means to do what Hollands (2008) asks of them. That is, 'under-gird' the social capital which is not only critical in underpinning their information systems and communication technologies in question, but also insightful in revealing the wider environmental and cultural role their networks of innovation and creative partnerships play in supporting the development of a community-led transition to smart cities.

The book also suggests these critical insights are equally significant for the reason they take Hollands' (2008) thinking full circle and in doing so offer the very alternative to the 'top-down' entrepreneurial-based business that logic called loud and hard for. An alternative that in this instance is grounded in the type of 'bottom-up' development which all of the chapters in the book offer examples of. Bottom-up developments which are aligned with the cybernetics of the social capital underlying the emergence of smart communities and what their networks of innovation and creative partnerships embed as the intelligence of such actions.

The clear governance message

This is the clear governance message that the book seeks to convey, irrespective of whether the chapters in question are either capturing the current state of the art (Deakin, Chapter 2), or offering the models by which to analyse the transition from intelligent (Paskaleva, Chapter 3; Deakin et al., Chapter 4; Komninos, Chapter 5) to smart cities (Deakin, Chapter 6; Paskaleva, Chapter 7). For it is evident that with all of the chapters, irrespective of whether they focus on the emerging state of the art or the modelling and analysis of the transition, these are the top-level governance issues to be bottomed out by way of the intelligence this embeds and through the open innovation systems they found. The open innovation systems that they found and the creativity this in turn draws upon as the pool of knowledge that institutes either a neo-evolutionary (Deakin and Leydesdorff, Chapter 8) or modified triple helix model of the transition (Deakin and Cruickshank, Chapter 9). Either a neo-evolutionary or modified triple helix model also supported by a set of smart city performance measures capable of analysing the transition (Caragliu et al., Chapter 10; Kourtit et al., Chapter 11).

Modelling and analysis

Equally clear is not just how many of the chapters share the same core reading of smart cities, but how many of the chapters also serve as a means to expand upon many of the key insights offered by the state-of-the-art surveys and single them out for further examination, modelling and analysis. Examples of this can be found in Paskaleva's survey of electronically enhanced governance (Chapter 3) and Deakin et al.'s examination of eGov service developments under the IntelCities community of practice (Chapter 4). Komninos' state of the art on intelligent cities (Chapter 5) and Deakin's statement on the embedded intelligence of smart cities offer another such example. Here Komninos states:

> It seems that we still lack a deeper understanding about what makes a city intelligent. We are still in the age of the digital, rather than intelligent, or smart cities. All definitions of intelligent/smart cities stress the use of information and communications technologies to make cities more innovative and efficient. But, they do not equally stress the need to understand the drivers of intelligence and the forms of integration among innovation actors, open, connected communities, their service applications, monitoring and measurement. (p. 92)

Deakin (Chapter 6) attempts to deepen this understanding by offering an extensive review of Mitchell's thesis on the transition from the city of bits (intelligent) to e-topia (smart cities). It goes very much against the grain, arguing that our current understanding of embedded intelligence and smart cities puts us on the verge of cultivating a new environmental determinism. An environmental determinism which this time around is cybernetic, based on the embedded intelligence of knowledge-based agents underpinning the networking of smart cities. To avoid repeating this mistake, attention is drawn to a more radical democratic, i.e. egalitarian and ecologically integral, account of intelligence and the opportunities this opens up for an emancipatory view of smart cities.

Echoing the concerns that Komninos raises about our understanding of the transition, Paskaleva (Chapter 7) states:

> to bring this complex agenda forward, new and consistent 'smart city' strategies are necessary, ones that can contribute to achieving urban sustainability and better quality of the life for the general citizen. With the advance of technologies, society's spirit of innovation is booming. Open innovation is emerging as the new paradigm for building the smart city. With it, government and developers can draw on the expertise, skills and knowledge of the citizens to develop the advanced services and goods that are relevant to the needs of the people and the urban environment. (p. 126)

Furthermore, Paskaleva also suggests that in order to overcome the criticisms Deakin sets out, there is a need: 'to move forward on to a new scale, a new logic, set of principles and agendas for smart cities'. This new scale, logic and principle, Paskaleva goes on to suggest, is something that can be realised if cities are smart in:

- Raising social interaction in the heart of the smart city model, in which the infrastructures and services are jointly and dynamically discovered, invoked and composed by providers and users alike.
- Creating open 'digital citizen-developer' communities and establishing private-public-people partnerships (PPPPs) to find dynamic and imaginative ways to interact and create, drawing inspiration and experience from open innovation and sustainable urban development.
- Building new collaborations and networks, so cities can understand innovation, innovators understand cities, citizens can become effectively engaged and users can become content and service producers and deliverers.
- Deploying convergent Future Internet platforms and services for the promotion of sustainable life and work styles in and across emergent networks of smart cities.
- Creating smart open innovation urban ecosystems – specific urban settings or innovation playgrounds which combine innovation and social and commercial activities to enable open innovation and showcase the benefits for localities of growing smarter and more sustainable.

The development of this agenda is taken up by Deakin and Leydesdorff in their chapter on the triple helix of smart cities. This chapter demonstrates how the triple helix model enables us to study the knowledge base of an urban economy in terms of civil society's support for the evolution of cities as key components of innovation systems. In this schema, cities are considered as densities in networks among at least three relevant dynamics: that is, in the intellectual capital of universities, in the industry of wealth creation, and in their participation in the democratic government which forms the rule of law in civil society. The effects of these interactions in turn generating spaces where ICTs are exploited to bootstrap the notion of creative cities as the knowledge base of intelligent cities and their augmentation into smart cities. Smart at exploiting ICTs that are not only creative, or intelligent in generating intellectual capital and creating wealth, but in the sense in which the selection environments governing their knowledge production make it possible for cities to become integral parts of emerging innovation systems. While the specific combination of knowledge products needed for these sub-dynamics to align with one another and allow ICTs to bootstrap the notion of creative cities is not given, it is proposed the very reflexive instability of the intellectual capital and wealth creation wrapped up in this innovation system is what the co-evolutionary mechanism offers them to be smart in governing the globalisation of their 'next-order' dynamics.

As they go on to state:

> The capacity to process this transition reflexively, that is, in terms of translations [from creative, to intelligent and as part of the transition to smart cities], marks this a development which takes us beyond the dismantling

of national systems and the construction of regional advantages. Using this neo-evolutionary perspective of the triple helix model, it can be appreciated that cultural development, however liberal and potentially free, is not a spontaneous product of market economies, but the outcome of policies, academic leadership qualities, and corporate strategies, all of which need to be carefully reconstructed, pieced together and articulated before management can govern over them. (p. 146)

Deakin and Cruickshank's chapter on SCRAN has taken this representation of the triple helix further and gone on to offer an operational model of this evolutionary perspective on smart cities. As such the network serves to demonstrate the worth of the policies, academic leadership qualities and corporate strategies, which have been carefully reconstructed, pieced together and articulated so that management can govern over the very transition they oversee. In other words offer a worked example of how this particular translation of the knowledge products, and the dynamic it captures and represents, provides the means to cultivate those selection environments that give cities the opportunity to be smart in building the capacity that is needed for this regional innovation system to stabilise the transition as part of a transnational strategy.

In many ways both these chapters do what Paskaleva asks of smart cities, in the sense they get beyond the more anthropocentric driven logic of the living lab experiments and an obsession with the Future Internet, by underscoring the social and environmental significance of such technologies and by then going on to highlight the cultural significance of smart cities as key driving forces within urban and regional innovation systems. Within urban and regional innovation systems, it might also be added, whose scope, scale and reach are of transnational significance in terms of the developments they give rise to and the process of global change they share.

The transnational significance of such development and of the global change they signal is perhaps best documented in Caragliu et al.'s chapter. Their definition of smart cities and analysis of such development lead them to suggest that the change:

> Show[s] consistent evidence of a positive association between urban wealth and the presence of a vast number of creative professionals, a high score in a multimodal accessibility indicator, the quality of urban transportation networks, the diffusion of ICTs (most noticeably in the e-government industry) and, finally, the quality of human capital. These positive associations clearly define a policy agenda for smart cities, although clarity does not necessarily imply ease of implementation. (p. 186)

This, they suggest, is because:

> All variables shown to be positively associated with urban growth can be conceived of as stocks of capital; they are accumulated over time and

are subject to decay processes. Hence, educating people is on average successful only when investment in education is carried out over a long period with a stable flow of resources; transportation networks must be constantly updated to keep up with other fast-growing cities, in order to keep attracting people and ideas; the fast pace of innovation in the ICT industry calls for a continuous and deep restructuring and rethinking of the communication infrastructure. (Ibid.)

Echoing the contributions from Komninos, Deakin, Paskaleva, Deakin and Leydesdorff and also Deakin and Cruickshank, Caragliu et al.'s chapter also goes on to suggest that if policies towards smart cities are going to be successful in maintaining the types of positive associations they have hitherto been assumed to harbour, there shall not only need to be a deep restructuring of the ICT sector, but complete rethinking of the communication infrastructure.

This evidence base for such a deep restructuring and complete rethinking is what also underlies the final contribution to this book. As an advanced triple helix and network for framing the performance of smart cities, this deep restructuring and complete rethink takes the triple helix model of smart cities set out by Deakin and Leydesdorff and translated by Deakin and Cruickshank as a point of departure. Reviewing the metrics of smart cities, it serves to reiterate many of the debates found in the previous contributions and echoes their findings by:

> adding another unifying factor to the analysis, namely urban environments and their contour conditions. While it is accepted by the authors of this chapter that knowledge is created by the interplay of the relations captured in the original triple helix, with the advanced triple helix model it is proposed that the accumulation of capital within regional innovation systems is enhanced by way of interaction with urban environments and through their contour conditions. Contour conditions that not only contribute to the generation of intellectual capital, or to the creation of wealth within smart cities, but also to the standards that government draws upon to regulate the accumulation of capital as wealth within regional innovation systems. (p. 200)

This, the contributors suggest is important because:

> The results obtained stress the importance of analysing several dimensions of the urban environment in order to assess smart city performance and link it to the elements of the advanced triple helix. This is the reason why we have developed a further analytical step, considering four clusters of indicators, linked together and connected to one another in a network model. Here an analytical network framework has to be adopted, as it helps capture the triple helix of a smart urban development and verify whether the implied transformation of cities is based not merely on an

index of intellectual capital, but on a measure of wealth creation whose standards of governance are set in the public realm. (p. 213)

Meeting the challenge

It is suggested that, together, the contributions making up this book do not just manage to stake out a common ground for them to relate to one another, but also go a considerable way to meet the challenge Hollands (2008) sets when he asks the real smart city to stand up.

While a relatively simple thing to ask, the book goes some way to illustrate the true significance of what is being asked by Hollands (2008). For in order to get beyond the rhetoric of cities that claim to be smart, it has been necessary for those keen in uncovering the truth of the matter to not only survey the status of the cities proclaiming to be smart, but assemble the platform by which to do so. That platform, which in this instance necessitates the contributors to the smart city debate assembling a whole raft of instruments. Instruments that include the models, networks, analytical frameworks and metrics that make it possible to measure their performance as smart cities. Models and networks that have not offered any quick fixes, but instead have needed to be worked-up as the analytical frameworks by which metrics can be applied to meet the performance measurement challenge smart cities pose.

Putting the problem of instrumentalisation aside, the findings generated do clearly go some way to allay many of the fears Hollands (2008) originally raised about the emergence of smart cities. For as has clearly been demonstrated, none of the contributions to this book propagates the image of the smart city as little more than a set of technical issues which surround the entrepreneurial-based and market-driven logic of business models. On the contrary, they instead go to extreme lengths to do just the opposite and highlight what are referred to as the information-rich and highly communicative qualities of the technologies supporting them.

Lingering concerns

Putting the symbolism of this information-rich and highly communicative representation of smart cities aside, perhaps the main concerns that linger over their development relate back to the degree to which the attending innovation systems are either fully open, and therefore socially inclusive, or partially closed off and, as a consequence, exclusive in both cultural and environmental terms. That is to say, rest on innovation systems whose reinvention of cities is not so smart. Not so smart because they either generate social enclaves, or create gated communities and are therefore insufficiently restorative to be socially equitable and environmentally just in producing the type of knowledge smart cities need to meet the sustainable economic growth requirement.

The contributions from Komninos, Deakin, Paskaleva, Deakin and Cruickshank, Caragliu et al. and Kourtit et al. all make reference to this in terms of whether the cultural and environmental significance of the emerging innovation systems shall be as a means to either reproduce the status quo or if they will merely serve to exacerbate the divisions already deeply engrained within civil society. As with Deakin, Paskaleva's concern lies with the adverse effects that any such fault line within the system will have on low-income communities already caught in the digital divide. In particular, the adverse effect of doing little more than systematically reproducing the status quo by either blocking their participation in the emerging innovation system, or locking them out of the means by which to access the wealth they create. Wealth they would otherwise appropriate, both by way of and through the means deployed to govern any such distribution.

Based on this, it is evident that while the contributors have done much to allay the fears surrounding the entrepreneurial-driven logic of the business models supporting the transition, anxieties about the social intelligence of the information-rich and highly communicative qualities of the technologies supporting the transition still persist. For it appears that the degree to which the accumulation of social intelligence and the deployment of such capital in the creation of wealth can be relied upon to undercut the market economics of entrepreneurial-driven business models is a matter that many of the contributors still consider to remain unresolved.

This is perhaps why Paskaleva calls for the emerging innovation system to foster:

> new bottom up approaches based on user-generated content, social media and Web 2.0 applications [that open] up vast possibilities for a new interpretation and understanding of spatial differences and local effects, seen through the experiences of the citizens themselves, these in turn leading to new forms of citizen empowerment. (p. 117)

New forms of citizen empowerment which have:

> the potential for citizens to build not just the social capital, but the capacities required to become co-creators and co-producers of new and innovative services with the means to ensure that they are delivered in more effective and inclusive ways, taking full advantage of new internet-based technologies and applications. (Ibid.)

However, given the absence of any methodology to support the call for citizen-led co-creation and co-production, statements about the value of such innovations probably work best to highlight the true nature of the challenge this poses for those wanting to integrate such features into innovation systems. For as meta-stabilisation mechanisms such statements tend to be overloaded with normative intent and are unable to reveal where the integration of any such innovations can systematically open up the spaces needed for smart cities to turn things around. Despite all of their ground-breaking features, calls for co-generation and co-production don't currently offer a methodology to integrate such innovations into pre-existing systems.

In the absence of any such logic, they currently take on the status of meta-narratives, lacking not only the principles, but also the intermediate concepts needed for the intelligence they hold to systematically evolve as innovations. In particular, and in this instance, as innovations not only able to create wealth, but also capable of cultivating the environments that communities need to construct the type of citizen-led change that they call for. To achieve this it is not so much agendas but models that are needed. Moreover, models which are able to systematically capture the change in question, and how the system can evolve in response to this change and innovate as part of a creative act.

These needs and requirements are drawn attention to because without them there is a real danger the next question we shall be asked to answer is: will the real smart city model please stand up and be counted! Why? Well, not because anything said about the transition doesn't make any sense and therefore doesn't stand up to scrutiny, but for the reason it adds up to something quite the opposite. That is, the real need for a formally explicit scrutiny of the reasoning, concepts and measures they are founded on and which are still left concealed in the innovations systems that have been drawn attention to. This results in an unfortunate situation whereby the institutions (universities, industry and government, respectively) and communities (scientific, technical, professional and lay, situated either in the corporate sector or in the public realm) they also support are unable to do anything but fall back into the type of Mode 2 thinking currently in existence. That mode of thinking, trans-disciplinary, scientific and technological logic which is instrumental in promoting the type of unreconstructed cultural identity thinking and environmental determinism whose failing logic, line of reason, let alone policy limitations, we have already drawn attention to. Not only in the sense they have to be seen as culpable in merely reproducing the status quo, but also instrumental in exacerbating the social divisions, cultural nihilism and environmental destruction also attending the somewhat less than clever actions cities are just as equally accountable for.

The call for policy as a strategy to supplement a people-based and user-centric open innovation system doesn't really resolve the situation either. For rather than being explicit about the reasoning, concepts and measures they are founded on, the highly intuitive nature of such statements still conceals how the policies will build on the intelligence of these bottom-up, user-generated cultures and be smart in supporting environments capable of delivering the 'of the people and or the people', namely democratic solutions, they call for. Indeed the call for innovation to be systematically creative in not only cultivating links to the environment, but connections to urban policy, surfaces more as a proposal for this new coupling to set the contributions of university and industry (i.e. of intellectual capital and wealth creation) aside. Set them aside and concentrate instead on how urban policy can cultivate a top-to-bottom rewrite of the rule book for smart cities based on the environmental governance of the Future Internet.

Unfortunately, there does not seem to be any consensuses on how to construct the recombinant spaces of this rule book. For if we compare the positions taken by Komninos and Paskaleva (Chapters 5 and 7), we are told by the former that this calls for a deeper understanding of the intelligence that such a hyper-creative act generates, and by the latter, a greater knowledge of the environmentally sensitive urban policies needed for the attending user-developers to govern this transition towards smart cities. Indeed, while the former calls for an explicit link to be formed between them, the latter is content for the coupling to be loose. So loose as to leave the whole question of intellectual capital, and wealth creation for that matter, cut off from culture altogether as part of their search for a closer connection to the environment. This line of thinking might be said to bend the stick too far back in favour of Landry's (2008) 'creative', as opposed to Komninos'(2002, 2008) 'intelligent' city, and in favour of a situation whereby it is the former's information systems and the latter's communication technologies which become the basis for and driving force behind what Deakin and Al Wear (2012) refer to as the 'transition from intelligent to smart cities'.

This is perhaps why Deakin, Deakin and Leydesdorff, Deakin and Cruickshank and Caragliu et al. all find such intelligent, versus cultural and environmental, accounts of smart cities less than constructive. This may also go some way to explain why Deakin and Leydesdorff, Deakin and Cruickshank and Kourtit et al. all go on to advance an alternative based on knowledge production. An alternative whose solution corridor does much to set out the triple helix of knowledge production, along with the institutions, form and content this not only embeds as the intelligence of smart cities, but goes on to network as the wealth creation of their regional innovation systems. For unlike those who see smart cities as an index of the Future Internet, the alternative these contributions advance does not do this, either in terms of the intelligence, culture or environment they rest on, but in relation to the social dynamics of their innovation systems. The social dynamics of innovation systems, it might be added, that co-evolve. Not only by way of networks which the intellectual capital and wealth creation of the triple helix is founded on, but through a cultural reconstruction. In particular, a cultural reconstruction whose reinvention of cities as smart is significant for the reason the environment this creates in turn supports a regional innovation system capable of setting the standards for governing the information systems and communication technologies making up the Future Internet.

Otherwise, the tendency there is to reduce these interaction effects among the intellectual capital and wealth creation of respective governance regimes, to one contextualising the other as the mere conditions of their existence, leads the emerging Mode 2 discourse on smart cities either towards an economic representation of innovation systems, or to a singularly one-sided account of their scientific and technical qualities. What they both have in common is the tendency to take on forms that are unable to unpack the contents of the socially organised learning processes that smart cities are understood to not only cultivate, but provide the environment for. This is because within trans-disciplinary accounts of this kind, both the representations of innovation systems and the products of science and technology are absorbed as exogenous manna from heaven, which in turn means the process of absorption that is fostered by way of their associated polices, leadership and corporate strategies cannot be improved on other than by adaptation and imitation. From the vantage point of a neo-evolutionary perspective and triple helix model, however, these predominantly social situations and the communities they in turn cultivate can be hypothesised as relevant selection environments. The codes operating in these selection environments can then in turn be reconstructed, adjusted and strengthened by way of their interaction with local settings and through the lessons learnt about how to translate the knowledge forming the content of the communication.

This representation of smart cities differs markedly from other accounts of their emergence. For with these accounts, the focus of attention is not so much on what smart cities reveal about the social dynamic of organised knowledge production, either in terms of the meta-stabilising tendencies of their cultural reconstruction, or of the selection environments that this offers, but with discovering a typology that offers a critical insight into the Future Internet as the index of such developments. Unfortunately, the tendency these studies have to wrap the evolution of smart cities around a typology of innovation in information systems and communication technologies not only misses the point, but goes very much against the grain of the ongoing debate. For while all of them state that any such investigation into what is smart should begin with the city, they tend to ignore such statements and go on instead to study the instrumentalisation of their information systems and communication technologies. The problem with these accounts of smart cities is that in setting the city aside and concentrating instead on the ambient intelligence, culture, environments and ubiquitous computing of their information systems and communication technologies, they still end up facing the wrong way. That is, away from policy-relevant questions about whether the cultivation of such environments actually does anything to bridge the digital divide and if the ubiquity of such computing, their information systems and communication technologies is socially inclusive, let alone equitable. The explanation for these apparent inconsistencies lies in the unquestioning assumptions they make about the value of the intelligent city and the ongoing contribution that culture and environment can make to their accounts of what is particularly smart about them.

The smart camp emerging around the Future Internet should take note of the more recent accounts of the intelligent city. For their typology is not only founded on this notion of the city, i.e. as an augmented space where the virtual triumphs over the physical, the electronically enhanced replaces the face-to-face and this in turn results in the situation whereby services are delivered over the web, rather than the counter, but also rests on an equally well articulated account of what is smart about the culture of these environments. Unfortunately those advocating the virtues of the Future Internet account of smart cities are unable to see the value of what the intelligent city legacy contributes to the smart debate. The reason for this is simple, but a little disconcerting. This is because those advancing the Future Internet as the all-pervasive technologies of next-generation information-based communications are still locked into the entrepreneurial-driven logic of the market economics and corporate bureaucracy they may well oppose, but whose culture and environment their accounts of smart cities are still unable to transcend.

Such a lock-in results because these accounts of smart cities are unable to embed the types of networked intelligence needed for any innovation to push information systems beyond the market economics of corporate bureaucracy and towards the communications required to pull cities up onto a platform which can be smart in anything but technical terms. The error made in this regard is to assume that ideas emerging about the content-driven push of Web 2.0 platforms, with 4G cables and wireless sensors supporting the Internet of Things, can spin a web strong enough for any such network to set the stage for a multitude of civic values to evolve which are socially inclusive, let alone equitable. The origins of this error can be traced back to the purely technical, as opposed to institutional, concept of civil society they work with and offer as a public realm. For with the former, civil society is seen as yet another institution of state that stands alongside the other three pillars, whereas with the latter, universities, industry and government are understood to be institutions of civil society and public realm upon which their legal-democratic constitution rests.

Such errors suggest that the typology of smart cities as an index of the Future Internet is of limited value and something in excess of this is needed to avoid the mistake of assuming cities can be separated from their information systems and communication technologies without the social status of the former being compromised by what is claimed to be smart about the latter. This in turn suggests that if we are to avoid any such a misunderstanding, then:

- something in excess of a typology is needed to account for what is known to be smart about cities;
- the excess in question needs to extricate itself from the legacy of intelligent versus cultural and environmental accounts of smart cities;
- this requires that we do not step back into the so-called creative city, but recognise the significant opportunity the debate over intelligent cities offers us to overcome the current divisions between the intelligent, cultural and environmental representations of smart cities;
- this means we have to replace the extensively neo-conservative brand of thinking which is emerging to represent what is smart about cities, with

a platform which does not fall back on to either the transactional logic of market economics, or the corporate bureaucracy of the nation state, but moves forwards into the embedded intelligence, culture and environment of their regional innovation system;

•   any such knowledge of what is smart about cities has to move forwards in the direction of a regional innovation system. In particular, a regional innovation system which has the embedded intelligence, culture and environment capable of constructing a platform that offers the type of radically progressive liberal-democratic alternative called for.

The excess of such a regional innovation system, the multitude of civic values, be they in the form of intellectual capital, wealth creation or standards of governance, that it gives rise to and the communities that, perhaps more importantly, contain the practical instantiations of the policies, academic leadership and corporate strategies they not only carefully construct, but cultivate, piece together and articulate as environments, is what substantiates smart cities. In particular, as a well grounded and principled model which is able to demonstrate the worth of the policies, leadership and corporate strategies, themselves carefully constructed, pieced together and articulated, as the means to cultivate the types of environments offering cities the opportunity for their information systems and communication technologies to be smart in creating wealth from a process of customisation. From a process of customisation whose multitude of civic values not only underpin the co-design of services, but construct the user profiles which are needed for the governance of such a regional innovation system to support the transition as part of a transnational strategy.

State of the transition

Reflecting on the emergence of smart cities, this book has drawn upon the triple helix as the means to advance a critique of recent accounts offered to explain their development. It has sought to explain the emergence of smart cities in terms of the all-pervasive technologies of information-based communications currently being exploited to generate the notion of creative cities, as the knowledge base of intelligent cities and their augmentation into smart cities. Cities that are smart because their ICTs are clever in embedding the intelligence needed to not only generate capital, or create wealth, but to also cultivate the selection environments that govern their knowledge production and make it possible for them to become integral parts of emerging regional innovation systems.

Based on this understanding, pre-existing accounts of smart cities have been subjected to a critical transformation. The triple helix that emerges, the institutions that the resulting model embodies, the form and content they in turn relate to not just socially, but also in cultural and environmental terms, have then been traced through the communities this networks as part of an evolving regional innovation system. This has drawn attention to the net-working of the triple helix as a model of knowledge production.

The book has also argued that the reflexive instability of such a knowledge-based system provides the co-evolutionary mechanism for an informational, technological and communicative meta-stabilisation, and one which offers society the possibility of being smart in relating this development to their 'next-order' dynamics. The capacity the triple helix has to process this transition reflexively; that is, in terms of translations, captures a dynamic which takes us beyond the dismantling of national systems and the construction of regional advantages. Likewise it has been argued the social dynamic of such disorganising and fragmenting trajectories can also be turned in the direction of the local, so that cities can take advantage of, be clever, namely smart in deploying information and communication technologies capable of meeting the needs of the excluded and the inequities that such a meta-stabilisation requires to regulate.

It has also been suggested this dynamic is significant because it begins to throw some light on what is smart about the capacity-building exercises cit-ies are entering into. That is to say, reveal that what is smart does not rest so much with the technologies that support cities, as with the generation of intel-lectual capital, the creation of wealth and regulation, which underpins their application. In particular, the intellectual capital whose experiential learning provides communities with the capacity to build such a platform and then go on to exploit the ambient intelligence such a process of wealth creation gen-erates as smart environments. As smart environments that go on to support the co-design, monitoring and evaluation of electronically enhanced services, developed by way of and through the ubiquitous computing of the multi-channelled access that cities offer to given user profiles.

This critical insight suggests it is the intellectual capital of the experiential learning tied to this knowledge base which provides an index of how smart cities are in openly sourcing their ambient intelligence and the wealth-creating opportunities technologies offer, be they the ICTs of Web 2.0 services or the Future Internet of Things, to be smart in constructing environments support-ing their application. Smart in constructing environments that support their application, it might be added, as predominantly social as opposed to techni-cal experiences. If only because the intellectual capital of the knowledge base and experiential learning that such exercises in ubiquitous computing help create are not only collective in their form (ambient intelligence), but openly sourced (as opposed to privately appropriated) by members of the public. Openly sourced by members of the public and collectively appropriated by communities, it might also be added, rather than corporate bureaucracies, as the curators of the content, proceeds and wealth of products emerging from the smart environments that cities are beginning to cultivate as the material basis for such technologies to democratise both their application and use.

This goes some way to bridge the gaps in previous representations of the model by using the networking possibilities that social capital offers to stabilise cities and secure their status as smart. That is, organise the intellectual capital of their learning communities to be smarter creators of wealth and the harbingers of standards within the public realm which govern over the knowledge production of regional innovation systems. The explanation for this lies in the realisation that in regional innovation systems such developments are networked not as technologies, but as social capital. In particular, as the social capital that is embedded in Web 2.0 technologies and in the ambition that their communities have to deploy them. That is to say, the ambition that they share to create wealth not gauged in terms of economic worth, but the creative value of their intellectual capital. The creative value of their intellectual capital that is significant because it generates the means by which the knowledge produced by these communities is no longer left as the tacit experiences of routine practices, but made explicit. Made explicit as codifications whose creative value rests with the capacity that their intellectual capital has to transfer knowledge of how communities can not only assemble the means to override economic interests, but mark themselves out culturally. Mark themselves out culturally for the reason they also offer sufficient power for this reconstruction and the environments it builds on behalf of the public in order to regulate the increasingly transnational status of such developments.

References

Deakin, M. and Al Wear, H., eds (2012) *From Intelligent to Smart Cities*, Abingdon, Routledge.

Hollands, R. (2008) Will the Real Smart City Stand Up?, *City*, 12 (3): 302–320.

Komninos, N. (2002) *Intelligent Cities: Innovation, Knowledge Systems and Digital Spaces*, London, Spon Press.

Komninos, N. (2008) *Intelligent Cities and the Globalisation of Innovation Networks*, London, Taylor & Francis.

Landry, C. (2008) *The Creative City*, London, Earthscan.

Index

References to figures and tables appear in bold.